Airborne and Air Assault Operations

March 2015

United States Government
US Army

Contents

	Page
PREFACE	ix
INTRODUCTION	x

Chapter 1	OVERVIEW	1-1
	Section I – Forcible Entry Operations	**1-1**
	Joint Principles for Forcible Entry Operations	1-2
	Operational Applications of Forcible Entry Operations	1-4
	Joint Command and Control	1-6
	Joint Operation Planning	1-13
	Joint Entry Force	1-16
	Section II – Vertical Envelopment	**1-18**
	Tactical Applications	1-18
	Command Responsibility	1-21
	Space Operation	1-25

PART ONE AIRBORNE OPERATIONS

Chapter 2	ORGANIZATION AND EMPLOYMENT	2-1
	Section I – Airborne Assault Force	**2-1**
	Organization of Forces	2-1
	Echelon Employment	2-2
	Section II – Airborne Assault Force Considerations	**2-4**
	Assault Force Formation	2-4
	Joint and Army Control Teams	2-4
	Section III – Capabilities, Limitations, Vulnerabilities	**2-5**
	Capabilities	2-5
	Limitations	2-5
	Vulnerabilities	2-6
	Section IV – Airborne Command and Control Platforms	**2-6**
	Airborne Warning And Control System	2-6
	Joint Surveillance Target Attack Radar System	2-7

Distribution Restriction: Approved for public release; distribution is unlimited.

This publication supersedes FM 90-26, dated 18 December 1990 and ATTP 3-18.12, dated 1 March 2011.

6 March 2015   FM 3-99   i

Contents

	Joint Airborne Communications Center/Command Post 2-7
Chapter 3	**AIRBORNE ASSAULT PLANNING** **3-1**
	Section I – Roles and Responsibilities... **3-1**
	Higher Headquarters.. 3-1
	Airborne Assault Force... 3-2
	Supporting Aviation 3-2
	Section II – Reverse Planning Sequence **3-2**
	Ground Tactical Plan.. 3-3
	Landing Plan 3-3
	Air Movement Plan 3-3
	Marshalling Plan... 3-4
	Section III – Planning Considerations **3-4**
	Planning Methodology.. 3-4
	Predeployment Planning and Preparation 3-6
	X-hour/N-hour Sequences for Deployment.. 3-6
	Optimize Available Planning Time 3-6
	Control Measures... 3-9
	Determine Go/No-Go Criteria 3-10
	Airfield Opening.. 3-10
	Section IV – Shaping Operations... **3-11**
	Create Conditions 3-11
	Preserve Conditions... 3-12
Chapter 4	**GROUND TACTICAL PLAN**... **4-1**
	Section I – Elements **4-1**
	Task Organization 4-1
	Mission Statement.. 4-1
	Commander's Intent... 4-2
	Concept of Operations 4-2
	Tasks to Subordinates 4-2
	Section II – Plan Development **4-3**
	Mission Variables of METT-TC 4-3
	Assault Objective and Airhead Line 4-5
	Section III – Air-ground Operations... **4-12**
	Fundamental Considerations 4-12
	Close Combat Attack.. 4-12
	Close Air Support 4-14
	Unmanned Aircraft System 4-15
	Section IV – Execution... **4-16**
	Conduct of the Airborne Assault 4-16
	Development of the Airhead... 4-17
	Buildup of Combat Power 4-17
	Section V – Follow-on Operations... **4-18**
	Section VI – Supporting Operations.. **4-18**
	Remote Marshalling 4-19
	Intermediate Staging Base... 4-20

Contents

Chapter 5	**LANDING PLAN**	**5-1**
	Section I – Delivery Considerations	5-1
	Organization	5-1
	Requirements	5-1
	Priorities	5-2
	Section II – Delivery Elements	5-2
	Sequence of Delivery	5-2
	Method of Delivery	5-3
	Place of Delivery	5-5
	Time of Delivery	5-9
	Section III – Preparation and Supporting Fires	5-9
	Fire Support Planning	5-10
	Fire Support Considerations	5-11
	Air Operations	5-14
	Section IV – Assembly and Reorganization	5-17
	Assembly	5-17
	Reorganization	5-25
Chapter 6	**AIR MOVEMENT PLAN**	**6-1**
	Section I – Joint Planning	6-1
	Section II – Elements of the Air Movement Plan	6-1
	Air Movement Table	6-1
	Types of Movement	6-2
	Aircraft Requirements	6-2
	Section III – Load Planning Considerations	6-3
	Tactical Integrity	6-3
	Cross Loading	6-3
	Self-Sufficiency	6-4
	Section IV – Loading and Delivery of Forces	6-5
	Load Planning Sequence	6-5
	Load Planning of Vehicles	6-5
	Air Movement Planning Worksheet	6-5
	Unit Aircraft Utilization Form	6-6
	Section V – Aircraft Load and Air Movement Table	6-6
	Section VI – Manifests and Air-Loading Planning System	6-9
	Manifests	6-9
	Integrated Computerized Deployment System	6-9
Chapter 7	**MARSHALLING PLAN**	**7-1**
	Section I – Preparation	7-1
	Section II – Movement	7-2
	Section III – Protection	7-2
	Passive Defense Measures	7-2
	Dispersal	7-3
	Section IV – Departure Airfield-Marshalling Area	7-3
	Selecting Departure Airfields	7-3
	Selecting and Operating Marshalling Areas	7-4

Contents

Facility Requirements 7-4
Marshalled Unit and Support Organization Activities 7-7
Section V – Outload 7-9
Outload Planning Considerations 7-9
Outload Control 7-10

PART TWO AIR ASSAULT OPERATIONS

Chapter 8   ORGANIZATION AND EMPLOYMENT 8-1
Section I – Air Assault and Air Movements 8-1
Section II – Air Assault Task Force 8-1
Organizing Forces 8-2
Brigade Combat Teams 8-2
Combat Aviation Brigades 8-2
Section III – Task Force Considerations 8-4
Section IV – Capabilities, Limitations, Vulnerabilities 8-5
Capabilities 8-5
Limitations 8-5
Vulnerabilities 8-6
Section V – Airspace Control 8-6
Concurrent Employment 8-6
Airspace Coordinating Measures 8-7
Airspace Development 8-8
Section VI – Air Assault Task Force Mission Command 8-10
Mission Orders 8-10
Command Posts 8-12
Personnel and Key Elements within the Task Force 8-13
Mission Command System 8-15

Chapter 9   AIR ASSAULT PLANNING 9-1
Section I – Roles and Responsibilities 9-1
Higher Headquarters 9-1
Brigade Combat Team 9-1
Supporting Aviation Units 9-1
Section II – Reverse Planning Sequence 9-1
Section III – Planning Methodology 9-2
Deliberate Planning 9-3
Time-Constrained Planning 9-4
Rapid Decisionmaking and Synchronization Process 9-5
Allowable Cargo Load Planning Considerations 9-6
Section IV – Planning Process 9-6
Warning Order 9-6
Initial Planning Conference 9-7
Air Mission Coordination Meeting 9-7
Air Mission Brief 9-8
Air Mission Brief Documents 9-8
Air Assault Task Force Rehearsal 9-9
Aircrew Brief 9-9

Contents

              Aviation Task Force Rehearsal 9-9
              Condition Checks 9-9
              Abort Criteria 9-10
            Section V – Control Measures 9-11
              Force-Oriented Control Measures 9-11
              Boundaries 9-12
              Fire Support Coordination Measures 9-12
              Airspace Coordinating Measures 9-12
            Section VI – Shaping Operations 9-13
            Section VII – Medical and Casualty Evacuation 9-14
              Medical Evacuation Planning 9-14
              Casualty Backhaul 9-15
              Medical Evacuation Landing Zone 9-15

Chapter 10   **GROUND TACTICAL PLAN 10-1**
            Section I – Elements 10-1
              Task Organization 10-1
              Mission Statement 10-1
              Commander's Intent 10-1
              Concept of Operations 10-2
              Tasks to Subordinate Units 10-4
            Section II – Plan Development 10-4
              Mission Analysis 10-4
              Assault Objective and Landing Zone Development 10-8
            Section III – Air-ground Operations 10-8
              Effective Integration 10-8
              Close Combat Attacks 10-9
              Close Air Support 10-11
              Unmanned Aircraft Systems 10-11
            Section IV – Execution 10-12
              Conduct of the Air Assault 10-12
              Buildup of Combat Power 10-12

Chapter 11   **LANDING PLAN 11-1**
            Section I – Landing Zone Selection 11-1
              Criteria for Selecting Landing Zones 11-1
              Location of Landing Zones 11-2
              Number of Landing Zones 11-2
            Section II – Landing Zone Updates 11-3
              Landing Zone Condition 11-3
              Fixed-wing Support 11-3
              Unmanned Aircraft System Support 11-3
            Section III – Hot Landing Zone Considerations 11-3
              Scenarios 11-3
              Reaction to Enemy Contact Away From the Objective 11-4
              Reaction to Enemy Contact on the Objective 11-4
            Section IV – Preparation and Supporting Fires 11-4

Contents

	Section V – Landing Site Operations	**11-5**
	Landing Zone and Obstacle Markings	11-5
	Exiting the Aircraft	11-6
	Exiting the Landing Zone	11-11
Chapter 12	**AIR MOVEMENT PLAN**	**12-1**
	Section I – Development Considerations	**12-1**
	Air Routes	12-1
	En Route Formations	12-3
	Terrain Flight Modes	12-5
	Fires	12-6
	Suppression of Enemy Air Defenses	12-6
	Air Assault Security	12-8
	Mission Command	12-8
	Section II – Air Movement Table	**12-9**
	Air Movement Table Development	12-9
	Air Movement Table Criteria	12-9
Chapter 13	**LOADING AND STAGING**	**13-1**
	Section I – Loading Plan	**13-1**
	Pickup Zone Selection	13-1
	Pickup Zone Organization and Control	13-2
	Coordination With Supporting Aviation Unit	13-3
	Preparation of Air Loading Tables	13-3
	Disposition of Loads on Pickup Zone	13-4
	Lifts, Serials, and Chalks	13-5
	Bump Plan	13-6
	Section II – Staging Plan	**13-7**
	Preparation for Loading	13-7
	Movement to Pickup Zone	13-7
	Chalk Check-In and Inspection	13-7
	Load Staging	13-8
	Sling Load Operations	13-8
GLOSSARY		**Glossary-1**
REFERENCES		**References-1**
INDEX		**Index-1**

Figures

Figure 1-1. Joint phasing model	1-5
Figure 3-1. MDMP and airborne assault planning process	3-5
Figure 4-1. Assault objectives	4-6
Figure 4-2. Airhead line	4-7
Figure 4-3. Boundaries	4-10

Figure 4-4. Base options 4-19
Figure 5-1. Offset and parallel drop zones 5-7
Figure 5-2. Parallel on-line drop zones 5-8
Figure 5-3. Assembly control posts for ABN IBCT forces landing on one drop zone 5-18
Figure 5-4. Movement of personnel to company assembly area 5-19
Figure 5-5. Line-of-flight/clock system 5-20
Figure 5-6. Stiner aid 5-21
Figure 6-1. Cross-loaded aircraft 6-7
Figure 7-1. Airborne task force marshalling area 7-5
Figure 7-2. Heavy-drop loading area control center 7-6
Figure 7-3. Heavy-drop rigging site 7-7
Figure 7-4. Concept of outload control 7-11
Figure 8-1. Grid line method 8-9
Figure 8-2. Attack by fire method 8-10
Figure 8-3. Example of air assault leadership positioning 8-12
Figure 9-1. Air assault planning stages 9-2
Figure 9-2. MDMP and air assault planning process 9-4
Figure 9-3. Time-constrained air assault planning 9-5
Figure 10-1. Organizational framework 10-3
Figure 10-2. Weather data 10-7
Figure 11-1. One-side off-load (UH-60) 11-7
Figure 11-2a. One-side off-load (squads in same chalk) trail landing formation 11-8
Figure 11-2b. One-side off-load (squads in same chalk) staggered trail right landing formation 11-8
Figure 11-3. Two-side off-load (UH-60) 11-9
Figure 11-4. Two-side off-load (squads in same chalk) diamond landing formation ... 11-10
Figure 11-5. Two-side off-load (chalks cross-loaded) heavy right landing formation .. 11-10
Figure 11-6. Rear ramp off-load and landing zone exit (CH-47) 11-11
Figure 11-7. One-side landing zone rush (squads in same chalk) trail landing formation 11-12
Figure 11-8. Two-side landing zone rush (chalks cross loaded) trail landing formation 11-13
Figure 11-9. Two-side landing zone rush (squads in same chalk) trail landing formation 11-14
Figure 12-1. Air route overlay 12-3
Figure 12-2. Standard flight and landing formations 12-5
Figure 13-1. Example pickup zone diagram 13-4
Figure 13-2. Lifts, serials, and chalks 13-6

Tables

Table 1-1. Command responsibility of airborne operations 1-22
Table 1-2. Command responsibility of air assault operations 1-24

Table 7-1. Parachute issue 7-2
Table 8-1. Heavy Combat Aviation Brigade organization 8-3
Table 8-2. Medium Combat Aviation Brigade organization..8-4
Table 8-3. Standard air assault radio networks and monitoring requirements.................... 8-18
Table 9-1. Example of an air mission coordination meeting agenda 9-8
Table 12-1. Example air movement table 12-10
Table 13-1. Example air loading table.. 13-3
Table 13-2. Aircraft bump information 13-6

Preface

Army Field Manual (FM) 3-99, *Airborne and Air Assault Operations,* establishes doctrine to govern the activities and performance of Army forces in forcible entry (specifically airborne and air assault operations) and provides the doctrinal basis for vertical envelopment and follow-on operations. This publication provides leaders with descriptive guidance on how Army forces conduct vertical envelopment within the simultaneous combination of offense, defense, and stability. These doctrinal principles are intended to be used as a guide and are not to be considered prescriptive.

FM 3-99 encompasses tactics for Army airborne and air assault operations and describes how commanders plan, prepare, and conduct airborne and air assault operations by means of joint combined arms operations. This publication supersedes FM 90-26, *Airborne Operations* and Army Tactics, Techniques, and Procedures 3-18.12, *Air Assault Operations*.

To comprehend the doctrine contained in this publication, readers must first understand the principles of war, the nature of unified land operations, and the links between the operational and tactical levels of war described in Joint Publication (JP) 3-0, Army Doctrine Publication (ADP) 3-0, and Army Doctrine Reference Publication (ADRP) 3-0. The reader must understand the fundamentals of the operations process found in ADP and ADRP 5-0 associated with the conduct of offensive and defensive tasks contained in FM 3-90-1 and reconnaissance, security, and tactical enabling tasks contained in FM 3-90-2. In addition the reader must also fully understand the principles of mission command as described in ADP 6-0 and ADRP 6-0 and command and staff organization and operations found in FM 6-0.

The principal audience for FM 3-99 is the commanders, staff, officers, and noncommissioned officers (NCOs) of the brigade, battalions, and companies within the brigade combat team. The audience also includes the United States Army Training and Doctrine Command institutions and components, and the United States Army Special Operations Command. It serves as an authoritative reference for personnel developing doctrine, materiel and force structure, institutional and unit training, and standard operating procedures (SOPs) for airborne or air assault operations.

Commanders, staffs, and subordinates ensure their decisions and actions comply with applicable United States, international, and, in some cases, host-nation laws and regulations. Commanders at all levels ensure their Soldiers operate in accordance with the law of war and the rules of engagement. (Refer to FM 27-10.)

FM 3-99 uses joint terms where applicable. Selected joint and Army terms and definitions appear in both the glossary and text. Terms for which FM 3-99 is the proponent publication (the authority) are marked with an asterisk(*) in the glossary. Terms and definitions for which FM 3-99 is the proponent publication are boldfaced in the text and the term is italicized. For other definitions shown in the text, the term is italicized and the number of the proponent publication follows the definition.

This publication applies to the Active Army, the Army National Guard/Army National Guard of the United States, and the United States Army Reserve unless otherwise stated.

The proponent for FM 3-99 is the U.S. Army Training and Doctrine Command (TRADOC). The preparing agency is the United States Army Maneuver Center of Excellence (MCoE). Send comments and recommendations by— mail or e-mail—using or following the format of DA Form 2028, *(Recommended Changes to Publications and Blank Forms)*. Point of contact information is as follows:

E-mail: usarmy.benning.mcoe.mbx.doctrine@mail.mil
Phone: COM 706-545-7114 or DSN 835-7114
Mail: Commanding General, Maneuver Center of Excellence, Directorate of Training and Doctrine, Doctrine and Collective Training Division, ATTN: ATZK-TDD, Fort Benning, GA 31905-5410

Unless otherwise stated, whenever the masculine gender is used, both men and women are implied.

This page intentionally left blank.

Introduction

Assured access is the desired conditions that the United States seeks to maintain in potential areas of strategic importance throughout the world. Assured access is the result of a combination of geopolitical factors that affect the United States' ability to exert influence and project power in a variety of political, economic, humanitarian, and military situations. While assured access includes the freedom of movement through the global commons of international airspace and oceanic sea lanes, operational access is the ability to project military force into an operational area with sufficient freedom of action to accomplish the mission.

Operational access is the joint force contribution to assured access, the unhindered national use of the global commons and select sovereign territory, waters, airspace and cyberspace. Operational access challenges can be categorized in a number of ways. They can be classified in terms of geographical, military, or diplomatic access issues. They can be described in terms of anti-access challenges, capabilities designed to prevent entry into an operational area, or area-denial capabilities designed to limit freedom of action within the operational area.

Army forces, as part of the joint force, contribute to projecting military force into an operational area and sustaining it in the face of armed opposition by defeating enemy anti-access and area denial capabilities. Projecting and sustaining forces require the capability to secure multiple entry points into an operational area. As a major application of forcible entry, the joint force uses vertical envelopment (airborne and air assault operations), giving leaders flexibility and depth to set conditions for decisive action. Vertical envelopment capitalizes on mobility to surprise the enemy, seize a lodgment, and gain the initiative.

The joint force establishes several precepts for gaining operational access. Foremost among them is "Conduct operations to gain access based on the requirements of the broader mission, while also designing subsequent operations to lessen access challenges." Consistent with this precept, entry operations into enemy territory are a means to military or political objectives, rather than an end in themselves. Entry operations are planned within the larger context of the campaign's overarching purpose. Without considering the impacts of entry operations on the larger objectives of a military campaign it is possible that entry may be gained through means that decrease the likelihood of achieving political and military objectives.

Operations during the last 25 years make clear that future entry of forces onto hostile or uncertain territory will be necessary for a number of purposes, one of which is the establishment of a lodgment. Joint doctrine defines forcible entry as the "seizing and holding of a lodgment in the face of armed opposition," and a lodgment as "a designated area in a hostile or potentially hostile operational area that, when seized and held, makes the continuous landing of troops and materiel possible and provides maneuver space for subsequent operations." Each service and functional component has an important role in how joint forcible entry operations overcome opposed access.

This manual provides doctrinal guidance for forcible entry operations at the operational and tactical levels. It describes relationships within the operational joint task force; vertical envelopment, organizational roles, functions, capabilities, and limitations; and responsibilities for the brigade combat team and its subordinate units within the assault force during airborne and air assault operations. A summary of key constructs to FM 3-99 follows:

- Operational access—forcible entry and vertical envelopment—within the operational environment.
- Task force organization, mission, capabilities, and limitations as well as the duties and responsibilities for the conduct of airborne and air assault operations.
- Task force command and staff operations; cross-functional staff organizations (cells, working groups, and centers) to assist in coordination.
- Meetings, working groups, and boards to integrate the staff, enhance planning, and decisionmaking within the task force.
- Airborne and air assault task force unique capabilities and planning considerations in transitioning to other tactical operations.

Chapter 1

Overview

Adaptive adversaries and enemies will contest United States joint forces across all domains— air, sea, land, space and cyberspace. Enemies are likely to employ anti-access strategies to prevent friendly force ability to project and sustain combat power into a region, and area denial strategies to constrain our nation's freedom of action within the region. Assured access— the unhindered national use of the global commons and selected sovereign territory, waters, airspace and cyberspace, is achieved by projecting all the elements of national power. Often the United States requires operational access— the ability to project military force into an operational area with sufficient freedom of action to accomplish the mission. Operational access is the desired condition that the United States seeks to maintain in areas of strategic importance, to achieve assured access. Army forces, as part of joint forces project forces into an operational area and conduct operations to defeat enemy anti-access and area denial capabilities and establish security conditions and control of territory to pressure freedom of movement and action for follow-on operations or deny that area's use to the enemy. Forcible entry operations are complex and always involve taking prudent risk to gain a position of relative advantage over the enemy. Equally critical is the transition between phases of the operation. This chapter discusses forcible entry operations and in particular the application of vertical envelopment as a tactical maneuver conducted by airborne or air assault forces to seize, retain, and exploit the initiative.

SECTION I – FORCIBLE ENTRY OPERATIONS

1-1. United States forces gain access to sovereign spaces through invitation (humanitarian relief), negotiations (basing rights), coercion (peacemaking operations under United Nation charter) or through force. The joint force's ability to project power and influence is challenged by proliferated anti-access weapons with increasing range, diversity, density and sophistication in the hands of both state and nonstate actors who are devising innovative approaches to contest joint forces in all domains. This evolving threat necessitates the development of comprehensive joint force solutions for gaining and maintaining operational access. (Refer to JP 3-18 for more information.)

1-2. The composite of the conditions, circumstances, and influences that make up an operational environment affects the employment of capabilities and impacts the decisions of commanders. The operational environment of the access force includes all enemy, adversarial, friendly, and neutral systems across the range of military operations; it includes an understanding of the physical environment, the state of governance, technology, local resources, and the culture of the local population. (Refer to JP 2-01.3 for more information.)

1-3. *Forcible entry* is the seizing and holding of a military lodgment in the face of armed opposition (JP 3-18). A *lodgment* is a designated area in a hostile or potentially hostile operational area that, when seized and held, makes the continuous landing of troops and materiel possible and provides maneuver space for subsequent operations (JP 3-18). A forcible entry operation is conducted to gain entry into the territory of an enemy by seizing a lodgment as rapidly as possible to enable the conduct of follow-on operations or conduct a singular operation. The operations must be designed to provide maneuver space for subsequent operations. This section addresses the principles and operational application of joint forcible entry

Chapter 1

operations. It discusses joint command and control, planning process and considerations, and the entry force.

JOINT PRINCIPLES FOR FORCIBLE ENTRY OPERATIONS

1-4. Fundamental principles are essential to plan, prepare, execute, and access joint forcible entry operations. Joint principles for forcible entry operations are:

ACHIEVE SURPRISE

1-5. Commanders and planners strive to achieve surprise regarding exact objectives, times, methods, and forces employed in forcible entry operations. Surprise depends upon comprehensive information-related capabilities [operations security (OPSEC) and military deception] followed by disciplined execution by the joint force. (Refer to JP 3-13.3 and JP 3-13.4 for more information.) Surprise is not a necessary condition for operational success (particularly when the force has overwhelming superiority), but it can reduce operational risk significantly.

CONTROL OF THE AIR

1-6. Counterair integrates offensive and defensive operations to attain and maintain a desired degree of air superiority and protection in the operational area to protect the force during periods of critical vulnerability and to preserve lines of communications. At a minimum, the joint force must neutralize the enemy's offensive air and missile capability and air defenses to achieve local air superiority and protection over the planned lodgment. The joint force controls the air through integrated and synchronized air and missile defense operations. Air interdiction of enemy forces throughout the operational area enhances the simultaneity and depth of the forcible entry operation.

CONTROL OF SPACE

1-7. Space superiority allows the joint force commander access to communications, weather, navigation, timing, remote sensing, and intelligence assets without prohibitive interference by the opposing force. Clearly defined command relationships are crucial for ensuring timely and effective execution of space operations and provide depth, persistence, and reach capabilities for commanders at the strategic, operational, and tactical levels.

ELECTROMAGNETIC SPECTRUM MANAGEMENT

1-8. *Electromagnetic spectrum management* is the planning, coordinating, and managing use of the electromagnetic spectrum through operational, engineering, and administrative procedures (JP 6-01). It includes the interrelated functions of frequency management, host-nation coordination, and joint spectrum interference resolution that together enable the planning, management, and execution of operations within the electromagnetic operational environment during all phases of military operations. The *electromagnetic operational environment* is the background electromagnetic environment and the friendly, neutral, and adversarial electromagnetic order of battle within the electromagnetic area of influence associated with a given operational area (JP 6-01). (Refer to JP 6-01 for more information.)

OPERATIONS IN THE INFORMATION ENVIRONMENT

1-9. *Information superiority* is the operational advantage derived from the ability to collect, process, and disseminate an uninterrupted flow of information while exploiting or denying an adversary's ability to do the same (JP 3-13). It enables the primary mission objective and information-related activities within information operations. *Information operations* is the integrated employment, during military operations, of information-related capabilities in concert with other lines of operation to influence, disrupt, corrupt, or usurp the decisionmaking of adversaries and potential adversaries while protecting our own (JP 3-13). (Refer to JP 3-13 for more information.)

1-10. Information in an operational environment is an important contributor to operational access as it enables commanders at all levels to make informed decisions on how best to apply combat power, ultimately creating opportunities to achieve decisive results. Operations in the *information environment*—the aggregate of individuals, organizations, and systems that collect, process, disseminate, or act on information (JP 3-13) requires—

- Complementary tasks of information operations that inform and influence a global audience and affect morale within the operational environment.
- Cyber electromagnetic activities (See FM 3-38.) to ensure information availability, protection, and delivery, as well as a means to deny, degrade, or disrupt the enemy's use of its command and control systems and other cyber capabilities.
- Knowledge management capabilities to allow the commanders to make informed; timely decisions despite the uncertainty of operations.
- Information management to help commanders make and disseminate effective decisions faster than the enemy can.
- Information systems to understand, visualize, describe, and direct operations.

SEA CONTROL

1-11. Local maritime superiority is required to project power ashore in support of the joint forcible entry operation and to protect sea lines of communications (SLOCs). Protection of SLOCs ensures the availability of logistic support required to sustain operations and support the transition to continuing operations by follow-on forces.

ISOLATE THE LODGMENT

1-12. A lodgment is a designated area in a hostile or potentially hostile operational area that, when seized and held, makes the continuous landing of troops and materiel possible and provides maneuver space for subsequent operations. The joint force attacks or neutralizes enemy capabilities with the potential to affect the establishment of the lodgment. These capabilities include—

- Enemy ground, sea, and air forces that can be committed to react to joint force assaults.
- Indirect fire systems and theater missile systems that can range the lodgment.
- Related enemy sensors, command and control systems, and digital networks.

GAIN AND MAINTAIN ACCESS

1-13. Gaining and maintaining access is a critical precondition for successful forcible entry and follow-on operations. In any given operational area, numerous and diverse limitations to access present themselves. Access may be restricted due to diplomatic, economic, military, or cultural factors. Ports, airfields, and infrastructures may be physically limited. Additional access precondition considerations include—

- Leveraging established basing, access, and security cooperation agreements as well as the regional and national expertise and partner capabilities developed through precrisis engagement activities at the national and regional levels.
- Appropriate shaping operations or activities focused on identifying and neutralizing an adversary's anti-access capabilities balanced against the need for surprise.
- Operational access to expand the degree to which the full range of joint capabilities within the joint operations area is utilized.

NEUTRALIZE ENEMY FORCES WITHIN THE LODGMENT

1-14. The joint force must neutralize enemy forces within the lodgment to facilitate the establishment of airheads (Refer to chapter 3 of this publication for more information.) and beachheads (Refer to JP 3-02 for more information.) within the operational area and to provide for the immediate protection of the force. Planning considerations should include—

- Identification of enemy infrastructure, which may be of value for future use by friendly forces.
- Limiting physical damage to lessen the time needed to rebuild.

Chapter 1

EXPAND THE LODGMENT

1-15. The joint force quickly builds combat power in order to enhance security and the ability to respond to enemy counter attacks, enable continuous landing of troops and materiel, and facilitate transition to subsequent operations. Analyze requirements to expand with regards to maximum on ground capabilities, throughput, and infrastructure.

MANAGE THE IMPACT OF ENVIRONMENTAL FACTORS

1-16. Managing the impact of environmental factors refers to overcoming the effect of land and sea obstacles; anticipating, preventing, detecting, and mitigating threat use of chemical, biological, radiological, and nuclear (CBRN), and weapons of mass destruction; and, determining the impact of climate, weather, and other naturally occurring hazards. (Refer to JP 3-11, JP 3-40, and JP 3-59 for more information.)

INTEGRATE SUPPORTING OPERATIONS

1-17. Reconnaissance, surveillance, security, and intelligence operations are critical to information collection based on the commander's critical information requirement. Information operations, civil-military operations, and special operations (to include special reconnaissance missions) are keys to setting conditions, and integrated into the operation at every stage from initial planning to transition. Logistic services comprise the support capabilities that collectively enable the joint force to rapidly provide sustainment of entry forces in order to achieve the envisioned end state of the joint force commander (JFC).

OPERATIONAL APPLICATIONS OF FORCIBLE ENTRY OPERATIONS

1-18. The Army combines campaign qualities and expeditionary capabilities to contribute decisive, sustained land power to unified actions. Campaign quality extends expeditionary capability well beyond deploying combined arms forces that are effective upon arrival. It is an ability to conduct sustained operations for as long as necessary, adapting to unpredictable and often profound changes in an operational environment (OE) as the campaign unfolds. Expeditionary capability is the ability to promptly deploy combined arms forces worldwide into operational environments (OEs) and conduct operations upon arrival. Future conflicts, involving forcible entry operations, place a premium on promptly deploying land power and constantly adapting to each campaign's unique circumstances as they occur and change.

CAMPAIGN QUALITY

1-19. The joint force commander (JFC) conducts campaigns to translate operational-level actions into strategic results and exploits the advantage of interdependent service capabilities. Through operational art and the principles of joint operations, the JFC determines the most effective and efficient methods for applying decisive operations in various locations across multiple echelons.

EXPEDITIONARY CAPABILITY

1-20. Expeditionary operations require the ability to deploy quickly with little notice, shape conditions in the operational area, operate immediately on arrival exploiting success and consolidating tactical and operational gains. Expeditionary capabilities of an entry force are more than physical attributes; they begin with a mindset that permeates the force. The Army provides entry forces to the joint force commander that are organized and equipped to rapidly deploy as well as conduct sustained operations.

PHASING AND TRANSITIONS

1-21. A *phase* is a planning and execution tool used to divide an operation in duration or activity (ADRP 3-0). Phasing is critical to arranging complex operations. It describes how the commander envisions the overall operation unfolding in time. Within a phase, a large portion of the force executes similar or mutually supporting activities. Achieving a specified condition or set of conditions typically marks the end of a phase. Descriptions of efforts during each phase should emphasize effort, concentrate combat power in time and space at a decisive point, and accomplish its objectives deliberately and logically.

1-22. Phasing is critical to arranging all tasks of an operation that cannot be conducted simultaneously. Commanders, with the assistance of the staff, visualize the mission, decide which tactics to use, and balance the tasks of unified land operations while preparing their intent and concept of operations (CONOPS). They determine which tasks the force can accomplish simultaneously, if phasing is required, what additional resources are necessary, and how to transition from one task to another. At the operational level, this requires looking beyond the current operation and prioritizing forces for the next phase or sequel.

1-23. Transitions mark a change of focus between phases or between the ongoing operation and execution of a branch or sequel. Transitions require planning and preparation well before their execution to maintain the momentum and tempo of operations. The force is vulnerable during transitions, and commanders establish clear conditions for their execution. Transition occurs for several reasons. It may occur from an operation dominated by combined arms maneuver to one dominated by wide area security. An unexpected change in conditions may require commanders to direct an abrupt transition between phases. In such cases, the overall composition of the force remains unchanged despite sudden changes in mission, task organization, and rules of engagement. Typically, task organization evolves to meet changing conditions; however, transition planning also must account for changes in mission. Commanders continuously assess the situation and task-organize and cycle their forces to retain the initiative. They strive to achieve changes in emphasis without incurring an operational pause.

1-24. The JFC's vision of how a campaign or operation should unfold and anticipated enemy action drives decisions regarding phasing. Generally, joint operations and campaigns involve six phases (shape, deter, seize initiative, dominate, stabilize, and enable civil authority) as illustrated in figure 1-1. Phasing assists in framing commander's intent and assigning tasks to subordinate commanders. By arranging operations and activities into phases, the joint force commander can better integrate and synchronize subordinate operations in time, space, and purpose. Each phase represents a natural subdivision of the campaign or operation. Within the context of the phases established by a JFC, subordinate JFCs and component commanders may establish additional phases that fit their CONOPS. A creditable threat of forcible entry operations can be an effective deterrence and may be applicable in both Phase 0 (Shape) and Phase I (Deter).

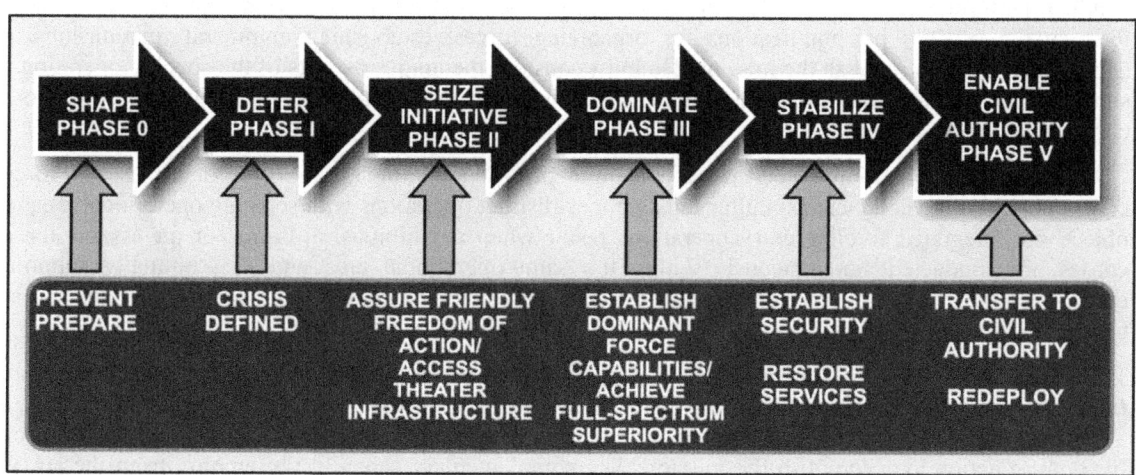

Figure 1-1. Joint phasing model

1-25. A forcible entry operation normally is conducted during Phase II (Seize the Initiative) or Phase III (Dominate) of a joint operation. A forcible entry operation may be the JFC's opening move to seize the initiative. For example, a JFC might direct friendly forces to conduct a vertical envelopment (airborne or air assault operation) to seize and hold a lodgment to ensure the continuous landing of troops and materiel and provide the maneuver space to conduct follow-on operations. The establishment of the lodgment, followed by the arrival and preparation of follow-on forces, usually marks the end of the forcible entry sub-phase of the operation and a transition to further offensive operations to seize the initiative or dominate.

Chapter 1

1-26. Forcible entry operations during the dominate phase of a campaign may be used for the following purposes:
- A sudden attack in force to achieve decisive results (a coup de main).
- Conducting operational movement and maneuver to attain positional advantage.
- A military deception.

1-27. The defeat of enemy forces usually marks the end of the dominate phase of the operation and a transition to area security and further operations to stabilize and enable civil authority. The mere existence of a forcible entry capability may be used by the JFC as a military deception operation, as a show of force or to force enemy movement even without mounting a forcible entry operation.

JOINT COMMAND AND CONTROL

1-28. Joint and partner interoperability (to include communications, planning and operations processes, staff functionality, language skills, and cultural knowledge) is critical to achieve unity of effort. Joint command and control information systems must enable interoperability and help synthesize information into knowledge while operating in austere environments, on the move, and across wide areas.

1-29. Joint command and control capabilities must maintain communications when networks are compromised or degraded due to friendly or enemy actions as well as materiel breakdown, natural atmospheric effects, or geospatial interference. The following paragraphs provide guidance on the employment options, organization of the operational area, command relationships, and command and control functions that support the conduct of forcible entry operations.

FORCE EMPLOYMENT

1-30. The combination of forcible entry capabilities employed depends on the mission. Unity of command is vital when multiple forcible entry capabilities are used or combined. Forcible entry operations are complex and must be kept simple in concept and well understood by all elements of the joint force and supporting commands.

1-31. If the JFC decides to use a combination of forcible entry capabilities, he must decide whether to conduct the forcible entries as concurrent or integrated operations. The distinction between concurrent and integrated operations has implications for organizing forces, establishing command relationships, and applying force to accomplish the mission. He must consider the unique aspects of the specific operation and should organize the force, establish command relationships, and apply force in a manner that fits the current situation.

1-32. Concurrent operations occur when a combination of amphibious, airborne, or air assault forcible entry operations are conducted simultaneously but as distinct operations with separate operational areas and objectives. Integrated forcible entry operations result when amphibious, airborne, or air assault forcible entries are conducted simultaneously within the same operational area and with mutually supporting objectives. Integrated forcible entry operations feature the complementary employment of forces and seek to maximize the capabilities of the respective forces available to the commander.

1-33. Dispersed joint forces use strategic and operational maneuver via air and sea to deploy or employ from the global system of main operating bases, forward operating sites, cooperative security locations, and amphibious and other sea-based platforms to project military force into an operational area with sufficient freedom of action to accomplish the mission. The complementary capability to employ from the air, with airborne and air assault forces, and the sea, with amphibious and air assault forces, complicates the adversary's defense. By requiring the adversary to defend a vast area against our mobility and deep power projection, joint forces can render some of his force irrelevant while exploiting the seams created in his defensive disposition.

1-34. Advanced force operations include strike operations, clandestine insertion of special operations forces and organic or supporting reconnaissance and/or surveillance teams, deception, counter-mine or counter-obstacle operations, and information operations. These activities combine with efforts to gain localized air and maritime superiority in the approaches to and entry areas. Joint strike operations along with theater air and missile defense, computer network operations, and electronic attack help provide protection and select areas of domain advantage allowing for the conduct of entry operations. (Refer to JP 3-05 and ADRP 3-05 for more information.)

1-35. To counter the enemy's defensive capabilities, joint forces attack from multiple directions and dimensions and disperse or concentrate forces depending on the intelligence preparation of the operational environment. (See JP 2-01.3.) Ideally, they avoid enemy defenses and seize undefended entry points through vertical envelopment, attacking the rear and flanks of a force, in effect cutting off or encircling the force. In hostile environments, assault forces conduct simultaneous force projection and sustainment to multiple unexpected or austere locations along a coastline and unexpected or austere landing zones ashore.

1-36. Operational maneuver requires the near-simultaneous movement and support of multiple tactical formations by ground, air, and sea from separate staging areas to locations in depth from which their military capabilities can be focused against critical enemy forces and facilities. High operational tempo and continuous pressure disrupts the enemy's ability to regroup, reconstitute capabilities, or reconfigure forces to support new plans. The primary means of maintaining continuous pressure is the rapid cycling of joint functions, related capabilities and activities grouped together to help JFCs integrate, synchronize, and direct joint operations.

1-37. An *intermediate staging base* is a tailorable, temporary location used for staging forces, sustainment and/or extraction into and out of an operational area (JP 3-35). Intermediate staging bases (ISBs) are a critical capability that enables shorter range recycling of intra-theater lift capabilities, reorganization and reconfiguration of capabilities to meet evolving tactical demands and recalibration of battle and logistical rhythms. Obtaining ISBs remains a critical part of baseline condition setting for joint force employment worldwide in access operations. Absent ISBs, the joint force's ability to maintain continuous pressure in the face of area denial capabilities is reduced significantly.

1-38. Joint commanders can employ forces in ISBs as a deterrent or as part of a forcible entry or other combat operation to defeat enemy forces. Besides conducting forcible entry to secure a lodgment as a prelude to a larger campaign, they may conduct limited attacks to seize key terrain upon which the enemy has emplaced air and missile defenses and anti-satellite missiles to ensure freedom of action in other domains.

1-39. Entry operations may rely on joint assets in an ISB for command and control, fires, protection, intelligence, reconnaissance, surveillance, and sustainment. Under austere conditions or when overcoming anti-access capabilities joint ISBs help reduce the requirement for large ground-based sustainment stocks and extended ground lines of communication, which could be vulnerable to enemy attack and which require additional forces to secure.

1-40. Forcible entry may include an airfield opening to support air operations following the initial entry. An *airfield* is an area prepared for the accommodation (including any buildings, installations, and equipment), landing, takeoff of aircraft. (JP 3-17). As airfields are identified for use and the missions and aircraft for which the airfield will be opened are determined; operators and planners generate options for one of three operational environments: permissive, uncertain, and hostile. Once employment forces are assigned by the joint task force, more detailed planning is conducted in coordination with the entry force and United States Air Force (USAF) first-responder airfield opening units.

ORGANIZATION OF THE FORCIBLE ENTRY OPERATIONAL AREA

1-41. Gaining operational area access. Joint forces seek to achieve periods of advantage in every domain to counter enemy anti-access capabilities and gain access to an operational area. They attack enemy anti-access capabilities and gain access to an operational area by conducting cross-domain operations in an effort to maximize the advantages and negate the disadvantages encountered in a single domain. Gaining total domain dominance is rarely obtainable, access forces fight for domain superiority—

- To gain an advantage in time and place that need not be permanent or widespread.
- At critical times and places within the operational area to achieve, the degree of freedom of action required to accomplish objectives.

1-42. Maintaining operational area access. The Army's role in the joint fight for maintaining operational area access is the integration of ground maneuver into a joint effort; conducting entry operations; and employing and sustaining forces while fighting to maintain freedom of action. The primary means of maintaining continuous pressure on the enemy is the continuous and rapid cycling of capabilities under operational-level direction throughout the duration of operations.

1-43. Operational area. Operational area is an overarching term encompassing more descriptive terms for geographic areas in which military operations are conducted. Operational areas include, but are not limited to, area of operations, amphibious objective area, joint operations area, and joint special operations area (JSOA). The JFC may designate operational areas on a temporary basis to facilitate the coordination, integration, and deconfliction between joint force components and supporting commands. Operational areas have physical dimensions comprised of some combination of air, land, and maritime domains and are defined by geographical boundaries.

1-44. Area of operations. Defined areas of operations for land and maritime forces typically do not encompass the entire operational area of the JFC, but should be large enough for the land and maritime component commanders to accomplish their missions and protect their forces. Component commanders with areas of operations may designate subordinate area of operations within which their subordinate forces operate. For example, the joint force land component commander (JFLCC) may assign subordinate commanders separate areas of operations within an assigned airborne or air assault area of operations. These subordinate commanders employ the full range of joint and service doctrinal control measures and graphics to delineate responsibilities, deconflict operations, safeguard friendly forces and civilians, and promote unity of effort.

1-45. Amphibious objective area. An amphibious objective area is an area of land, sea, and airspace, assigned by a joint force commander to commander, amphibious task force (CATF) to conduct amphibious operations. The amphibious objective area should be specified in the initiating directive. This area must be of sufficient size to ensure accomplishment of the amphibious force's mission and must provide sufficient area for conducting necessary sea, air, and land operations. This three-dimensional area often is limited in height (for example, up to 25,000 feet). Amphibious objective area air control procedures are identical to a high-density airspace control zone (HIDACZ) and CATF is the regional (or sector) air defense commander (RADC or SADC).The amphibious objective area extends below the water surface, and can have a depth limitation. (Refer to JP 3-02 for more information.)

1-46. Joint special operations area (JSOA). A joint special operations area is an area of land, sea, and airspace, assigned by a JFC to the commander of a joint special operations force to conduct special operations activities. The JFC may establish a JSOA when geographic boundaries between special operations forces and conventional forces are the most suitable control measures. Establishment of a JSOA for special operations forces to conduct operations provides a control measure and assists in the prevention of fratricide. The commander, joint special operations task force (CDRJSOTF) may request the establishment of a JSOA. When a JSOA is designated, the CDRJSOTF is the supported commander within the designated JSOA. The CDRJSOTF may further assign a specific area or sector within the JSOA to a subordinate commander for mission execution. The scope and duration of the special operations forces mission, operational environment, and politico-military considerations all influence the number, composition, and sequencing of special operations forces deployed into a JSOA. It may be limited in size to accommodate a discrete direct action mission or may be extensive enough to allow a continuing broad range of unconventional warfare operations. (Refer to JP 3-05 for more information.)

1-47. Airspace control area. An airspace control area is that airspace laterally defined by the boundaries that delineate the operational area. This airspace may include subareas. This airspace may entail an operational area and is a means of planning and dividing responsibility. While an operational area is in existence, airspace control within the operational area is delegated according to JFC guidance, the airspace control plan, and airspace control order. (Refer to JP 3-52 for more information.)

1-48. Control and coordination measures. Control and coordination of forcible entry operations pose a difficult challenge to all elements of the joint force. In addressing this challenge, the JFCs and appropriate commanders may employ various control and coordination measures that facilitate the execution of operations and, at the same time, protect the force to the greatest possible degree. These measures include, but are not limited to, boundaries that circumscribe operational areas; control measures to facilitate joint force maneuver; fire support coordination measures (FSCMs); and airspace coordinating measures. (Refer to JP 3-52 for more information.)

COMMAND RELATIONSHIPS FOR FORCIBLE ENTRY OPERATIONS

1-49. The JFC has full authority to assign missions, redirect efforts, and direct coordination between subordinate commanders to ensure unity of command. He may organize the forcible entry force as a subordinate joint task force or the forcible entry force may be organized from a component. An initiating directive provides guidance on command relationships and other pertinent instructions for the duration of the forcible entry operation. Joint force commander subordinate command relationships may include:

Functional Component Commanders

1-50. Designating joint force functional component commanders, allows resolution of joint issues at the functional component level and enhances component interaction at that level. Delegating control of the forcible entry operation to a functional component commander permits the JFC to focus on other responsibilities in the operational areas. Based on the JFC's guidance, the forcible entry operation may be conducted by a functional component commander. (Refer to JP 3-30, JP 3-32, and JP 3-02 for more information.) Responsibilities when organized under functional lines include the following:

- Joint force land component commander (JFLCC). Plans, coordinates, and employs designated forces or capabilities for joint land operations in support of the joint force commander's CONOPS. Normally commands forcible entry operations that involve airborne or air assaults that originate from land bases, and typically designates subordinate airborne and air assault task force commanders.
  - Airborne task force commander (ABNTFC). Serves as the airborne assault force commander's (ABNAFC) higher echelon commander. The ABNTFC may be the joint force commander or ground commander, depending upon experience and the scope of the operation. Responsible for the planning, coordination, and employment of designated airborne forcible entry forces or capabilities in the support of the joint force commander's CONOPS. The ABNTFC has overall responsibility to synchronize and integrate the actions of assigned, attached, and supporting air capabilities or forces in time, space, and purpose. The command relationships established between the joint force commander, ABNAFC, commander, airlift force; joint force air component commander (JFACC), and other designated commanders of the forcible entry force are key to mission success.
  - Air assault task force commander (AATFC). Serves as the overall commander of the air assault task force (AATF) using fixed- and rotary-wing aircraft deploying from land-based facilities and ships. Responsible for the planning, coordination, and employment of designated air assault forcible entry forces or capabilities in the support of the joint force commander's CONOPS. The AATFC ensures continuity of command throughout the operation by forming and employing the AATF, which is a temporary group of integrated forces tailored to a specific mission under the command of a single headquarters.
- Joint force air component commander (JFACC). Synchronizes and integrates the actions of assigned, attached, and supporting air capabilities or forces in time, space, and purpose in support of the joint force commander's CONOPS. The JFACC coordinates with the supported functional component commander or joint task force commander to establish airspace control and air defense plans in support of a forcible entry operation.
- Joint force maritime component commander (JFMCC). Plans, coordinates, and employs designated forces or capabilities for joint maritime operations in support of the joint force commander's CONOPS.
  - Commander, amphibious task force (CATF) and commander, landing force (CLF) amphibious assault forcible entry operations include air and land assaults that originate from the sea. The amphibious force is organized to best accomplish the mission based on the CONOPS.
  - Establishing command relationship between the CATF, CLF, and other designated commanders of the amphibious force is an important decision. An establishing directive is essential to ensure unity of effort within the amphibious force. Normally, a support relationship is established between the CATF and CLF by the JFC or establishing authority.

Special Operations Forces Command

1-51. When directed, Commander, United States (U.S.) Special Operations Command provides continental U.S.-based special operations forces to a JFC. The JFC normally exercises combatant command (command authority) of assigned and operational control of attached special operations forces through a commander, theater special operations command (TSOC), or a subunified commander. When a JFC establishes and employs multiple task forces concurrently, the TSOC commander may establish and employ multiple joint special operations task forces (JSOTFs) to manage special operations forces assets and accommodate joint task force/task force special operations requirements. Accordingly, the joint task force commander, as the common superior, establishes support or tactical control command relationships between the JSOTF commanders and joint task force/tasks force commanders. The special operations forces liaison to the JFC helps coordinate the operations of the supported or supporting special operations force and advises the joint force commander of special operations forces capabilities and limitations. (Refer to FM 6-05 for more information.)

Multinational Partners

1-52. Entry forces participating in a multinational operation always have at least two distinct chains of command: a national chain of command and a multinational chain of command. Although nations often participate in multinational operations, they rarely, if ever, relinquish national command of their forces.

1-53. Forcible entry operations with multinational partners are planned and conducted much the same as a U.S. joint force operation. Attaining unity of effort through unity of command for a multinational operation may not be politically feasible, but it should be a goal. Forcible entry objectives must be understood by all national forces. (Refer to FM 3-16 for more information.)

AIRSPACE CONTROL

1-54. The JFC normally designates a JFACC as the commander for joint air operations. In addition, to a JFACC, the JFC may designate an area air defense commander (AADC) and an airspace control authority or he may designate the JFACC as the AADC and airspace control authority. If a separate airspace control authority or AADC is designated, close coordination is essential for unity of effort.

Airspace Control Authority

1-55. The airspace control authority plans, coordinates, and develops airspace control procedures and operates the airspace control system. He monitors, assesses, and controls operational area airspace and directs changes according to the joint force commander's intent. The airspace control authority normally controls the airspace through the theater air control system (TACS) and the Army air-ground system (AAGS) in forcible entries. Situations may limit establishment of ground systems and require airborne or sea-based systems to conduct airspace control. Commanders and staffs should closely monitor and plan the employment of critical communication systems within TACS/AAGS. (Refer to JP 3-52 for more information.)

Army Air-Ground System

1-56. Army components of the AAGS consist of airspace elements, fire support cells, air and missile defense sections, and coordination and liaison elements embedded in Army command posts. Collectively they coordinate and integrate airspace use —joint, coalition, nonmilitary and Army manned and unmanned aircraft systems, directed energy, munitions— for the echelons they are assigned. These participants consist of airspace elements, fire support cells, air defense airspace management/brigade aviation elements (ADAM/BAEs), an Army air and missile defense command (AAMDC), battlefield coordination detachments (BCDs), ground and reconnaissance liaison detachments, and air defense artillery fire control officers (ADAFCOs). Some participants of the theater air-ground system (TAGS) —such as the air mobility liaison officer, the tactical air control party, and the air support operations center— remain under operational control of different Services but provide direct support during the conduct of operations. (Refer to FM 3-52 for more information.)

Airspace Elements by Echelon and Role

1-57. Airspace elements are organic to brigade combat teams BCTs and higher. Corps and division airspace elements are the same and both contain an airspace element in their main and tactical command posts. The BCTs contain a version of an airspace element referred to as an ADAM/BAE. The ADAM/BAE integrates brigade airspace, including air and missile defense (AMD) and aviation functions. Each of these elements coordinates with higher, subordinate, and adjacent elements to maximize the effectiveness of airspace control.

1-58. The ADAM/BAE manages the airspace control working group to facilitate and synchronize contributions from all the elements that perform the airspace collective tasks. The airspace control working group typically consists of an air liaison officer and representatives from the airspace element, aviation element, AMD element, fire support cell, tactical air control party, and unmanned aircraft systems element.

Airspace Coordination and Liaison Element

1-59. The JFACC establishes one or more joint air component coordination elements (JACCEs). JACCEs colocate with the joint force commanders headquarters and other component commanders' headquarters. When established, these elements act as the JFACC's primary representatives to the respective commanders and facilitate interaction among the respective staffs. The JACCE facilitates integration by exchanging current intelligence, operational data, and support requirements. It also aids integration by coordinating JFACC requirements for airspace coordinating measures (ACMs), FSCMs, close air support, air mobility, and space requirements.

1-60. The battlefield coordination detachment (BCD), the Army liaison to the JFACC, is located in the joint air operations center (JAOC). The BCD facilitates the synchronization of air and Army ground operations within the area of responsibility. BCD personnel work with their counterparts in the JAOC to facilitate planning, coordination, and execution of air-ground operations. The BCD expedites the exchange of information through face-to-face coordination and digital interfaces with JAOC elements and with—Army ground liaison officers at USAF operations centers, the Army theater main command post, the corps main command post (or if necessary the tactical command posts), and with subordinate unit command posts, if appropriate. At the corps or division main command post, the BCD exchanges information with the integrating and functional cells within the headquarters. (Refer to FM 3-52 for more information.)

1-61. The BCD also supervises the Army's reconnaissance liaison detachments and ground liaison detachments that provide coordination among Army forces and USAF reconnaissance, fighter, and airlift wings. The Army assigns ground liaison detachments to each USAF air wing operations center supporting ground operations. The Army ground liaison detachments provide Army expertise, interpreting and briefing pilots on the ground commander's concept of operations, tactics, equipment, and the ground situation. They also provide guidance on target designation, help identify friendly troops, and participate in the debriefing of pilots on their return from missions. These detachments are also the principal points of contact between the USAF contingency response groups and Army airfield control groups for controlling Army theater airlift movements. (Refer to JP 3-17 for more information on air mobility operations.)

Airspace Control Coordination During Airborne and Air Assault Force Operations

1-62. Missions such as airborne, air assault and other incursions into enemy territory require specific airspace control coordination. When supporting a forcible entry operation, the air component commander for the operation or JFACC (if designated) may use airborne command and control assets to enhance coordination and control of joint air operations and airspace management. Additional considerations include the following:
- Force employment and system interoperability normally determines the systems available to the airspace control authority in designating airspace control system to control joint air operations.
- Distances involved and the duration require establishing special air traffic control facilities or special tactics teams to extend detailed control into the objective area.
- Air traffic volume demands careful coordination to limit potential conflict and to enable the success of mission-essential operations within the airhead.

- Establishment of a high-density air control zone around a drop zone or landing zone (includes sufficient terrain and airspace) to permit safe and efficient air traffic control.
- A high-density air control zone can be nominated by the ground force commander and should include, at a minimum, the airspace bounded by the airhead line. (Refer to chapter 4 and chapter 10 of this publication for more information.)
- Within the high-density air control zone, all aircraft flights should be coordinated with the drop zone, landing zone, and the agency responsible for controlling the joint airspace.
- Air mission commander coordinates with the assault force commander to select the time on target and the direction of approach into and through the airhead.
- Environmental effects on airspace control coordination may dictate force employment techniques and aircraft selection for specific mission profiles.

AREA AIR DEFENSE COMMANDER

1-63. The JFC designates an area air defense commander (AADC) with the authority to plan, coordinate, and integrate overall joint force defensive counterair operations. The AADC normally is the component commander with the preponderance of air and missile defense assets and the capability to plan, coordinate, and execute integrated air and missile defense operations. Additionally, the AADC is granted the required command authority to deconflict and control engagements and to exercise real-time battle management.

1-64. As approved by the joint force commander, the AADC may designate the commander, Army air and missile defense command (AAMDC) as a deputy AADC for air and missile defense [DAADC (AMD)] in support of the AADC for defense counterair operations. The AAMDC is responsible for balancing the Army counterair assets/capabilities between the Army/JFLCC maneuver units and theater-level requirements. The AAMDC ensures that Army theater air and missile defense operations are internally coordinated and properly integrated with the joint force and multinational forces.

1-65. The Army provides mission command, sensors, and weapon systems for the counterair operational area, but does not provide the capability for regional or sector air defense commands within the land component area of operations. Regional or sector air defense commands normally are provided by command and control elements [control and reporting center (AADC), tactical air operations center United States Marine Corps (USMC), or Aegis United States Navy (USN)] of the other components.

1-66. The operational area, to include ingress and egress routes, must be fully protected by an integrated air defense system consisting of air, land, maritime, cyberspace, and space assets. The joint force is particularly vulnerable to attacks by enemy aircraft or surface-to-surface missiles during the early stages of a forcible entry. Accordingly, the primary objectives for air defense operations are to assist in gaining air superiority.

1-67. The AADC integrates all available surface-to-air assets into the overall air defense plan and complies with procedures and weapons control measures established by the joint force commander or JFACC. The AADC exercises the degree of control of all systems through established guidelines; weapons control status and joint force commander-approved procedural controls.

1-68. During air movement to the operational area, the AADC controls air defense operations from an airborne platform (for example, Airborne Warning and Control System). In practice, extended distances from staging bases to designated areas of operation may require the AADC to delegate control responsibilities to an air control element on board the airborne platform. Initial air defense assets may be limited to fighter aircraft only. Control of these aircraft is exercised through established procedural controls.

1-69. Forces initially entering the area of operation are accompanied by organic short-range air defense systems that must be integrated into the air defense architecture. Planned procedural control measures and guidelines may be established by the AADC to expedite integration of assets. With force buildup and the introduction of follow-on forces into the lodgment area, more robust high- to medium-altitude air defense systems normally become available. These systems must establish communications with the AADC's command and control agency and be incorporated into the established air defense system.

COMMUNICATIONS

1-70. Communications systems supporting forcible entry operations must be interoperable, agile, trusted, and shared. The complexity and tempo of assault force operations requires the technical capability to support an increased level of lateral coordination and integration between assault, special operations forces, and multinational forces in the operating area. Assault forces rely upon en route collaborative planning, rehearsal, execution and assessment tools and beyond-line-of-sight, over-the-horizon, on-the-move communications.

1-71. Typical forcible entry operations communications employ single and multichannel tactical satellites (TACSATs); commercial satellite communications (SATCOM); and single-channel ultrahigh frequency (UHF), very high frequency (VHF), and high frequency (HF) radios. When operating in degraded environments, assault forces must be prepared to operate using only line-of-sight or intermittent communications.

Communications System Planning

1-72. Communications system planning must be an integral part of joint force planning. Once specific command and control organization for the forcible entry operation is established, communications system planning begins and information exchange requirements are established.

1-73. The communication directorate of a joint staff (J-6) is responsible for planning and establishing the communications system and the communications estimate of supportability. The J-6 must be able to integrate communication across the joint force among elements conducting assault force operations, initial assault elements forward planning at an ISB, and the main assault force preparing for operations from home station.

1-74. Communications system planning is conducted in close coordination with the operations directorate of a joint staff (J-3) and intelligence directorate of a joint staff (J-2) to identify specialized equipment and dissemination requirements for some types of information. (Refer to JP 6-0 and FM 6-02 for more information.)

Communications Support During Airborne and Air Assault Force Operations

1-75. Communications requirements vary with the mission, size, composition, geography, and location of forcible entry forces and the senior headquarters. Significant considerations include the use of intermediate staging bases and airborne command and control platforms, to include en route mission planning and intelligence sharing, which can add to the complexity of managing the communications architecture.

1-76. Because communications systems must be built up at the objective area, some aspects of communications support are unique in forcible entry operations. Assault force support considerations include the following:
- Long-range radio communications through retransmission and relay sites with U.S.-based forces or intermediate staging bases to facilitate control of personnel, supplies, and equipment.
- Interoperability with the overall joint force communications architecture—communications redundancy for assault force and subordinate commanders.
- Initial deployment communications capability largely based on UHF SATCOM, becomes more robust as signal units and equipment enter the operational area.
- Command relationships, networks, frequency management, codes, navigational aids, and other communication issues must be resolved before the assault phase begins.

JOINT OPERATION PLANNING

1-77. Joint operation planning includes all activities that must be accomplished to plan for an anticipated operation: the mobilization, deployment, employment, sustainment, redeployment, and demobilization of joint forces. It integrates military power with other instruments of national power to achieve a desired military end state; the set of required conditions that defines achievement of the commander's objectives. Joint operation planning connects the strategic end state to campaign design and ultimately to tactical missions. The following paragraphs discuss the joint operation planning process and intelligence

Chapter 1

preparation of the operational environment, and planning considerations in support of forcible entry operations.

JOINT OPERATION PLANNING PROCESS

1-78. In conducting joint operation planning, commanders and staffs apply operational art to operational design using the joint operation planning process (JOPP). This includes forcible entry operation requirements. The JOPP is an orderly, analytical planning process comprised of a set of logical steps to analyze a mission; develop, analyze, and compare alternative course of actions; select the best course of action; and produce a plan or order. The process provides a methodical approach to planning at an organizational level before and during the joint operation. It focuses on the interaction between an organization's commander, staff, the commanders and staffs of the next higher and lower commands, and supporting commanders and their staffs.

1-79. Through the application of operational art and operational design and by using JOPP, the JFC and staff combine art and science (control) to develop products that describe how (ways) the joint force will employ its capabilities (means) to achieve the military end state (ends). Operational art is the application of creative imagination by commanders and staffs— supported by their skill, knowledge, and experience. Operational design is a process of iterative understanding and problem framing that supports commanders and staffs in their application of operational art with tools and a methodology to conceive of and construct viable approaches to operations and campaigns. Operational design results in the commander's operational approach, which broadly describes the actions the joint force needs to take to reach the end state. JOPP is an orderly, analytical process through which the JFC and staff translate the broad operational approach into detailed plans and orders. (Refer to JP 5-0 for more information.)

JOINT INTELLIGENCE PREPARATION OF THE OPERATIONAL ENVIRONMENT

1-80. The J-2 has the primary staff responsibility for planning, coordinating, and conducting the overall joint intelligence preparation of the operational environment (JIPOE) analysis and production effort at the joint force level. The JIPOE supports decisionmaking and planning by identifying, analyzing, and estimating the enemy's centers of gravity, critical factors, capabilities, limitations, requirements, vulnerabilities, intentions, and courses of action that are most likely to be encountered by the joint force.

1-81. Joint task force joint intelligence support element or joint task force joint intelligence operations center (JIOC) is the intelligence organization at the joint task force level that is responsible for complete air, space, ground, and maritime threat characteristics analysis for the joint operation. It identifies adversary centers of gravity; analyzes command, control, and communications systems, targeting support; collection management; and maintenance of a 24-hour watch in a full JIPOE effort.

1-82. The joint intelligence support element or JIOC continuously develops, updates, and tailors JIPOE products while proactively seeking out and exploiting all possible assistance from interagency and multinational intelligence sources. It directs the effort, integrating analyses with all products produced by subordinate commands and other organizations and ensures the JIPOE process encompasses a systematic analysis of all relevant aspects of the operational environment with tailored products continuously developed and updated to support the planning process. (Refer to JP 2-0 and JP 2-01.3 for more information.)

PLANNING CONSIDERATIONS IN SUPPORT OF FORCIBLE ENTRY OPERATIONS

1-83. Forcible entry, which may include airborne, air assault, and amphibious operations, or a combination of all three can create multiple dilemmas by creating threats that exceed the enemy's capability to respond. These operations are complex and high risk and should remain as simple as possible in concept.

1-84. Forcible entry operations require extensive intelligence, detailed coordination, innovation, and flexibility. Schemes of maneuver and coordination between forces need to be clearly understood by all participants. Forces are tailored for the mission and echeloned to permit simultaneous deployment and employment.

1-85. Entry forces require seamless intelligence and operations integration from the small unit up through national-level decision makers, enabling leaders at all levels to access relevant information at the proper time and place. The JIOC ensures the intelligence staffs of subordinate component commands have appropriate reconnaissance, surveillance, and intelligence products prepared for each domain in which entry forces operate.

1-86. As part of achieving decisive advantages early, joint force operations may be directed immediately against the enemy's center of gravity using conventional and special operations forces and capabilities. Attacks may be decisive or may begin offensive operations throughout the enemy's depth that can create dilemmas causing paralysis and destroying cohesion.

1-87. When airborne, air assault, and amphibious operations are combined, unity of command is vital. Rehearsals are a critical part of preparation for forcible entry. Participating forces need to be prepared to fight immediately upon arrival and require robust communications and intelligence capabilities to move with forward elements. The forcible entry force must be prepared to immediately transition to follow-on operations, and should plan accordingly.

1-88. Force entry actions occur in both singular and multiple operations. These actions include establishing forward presence, preparing the operational area, opening entry points, establishing and sustaining access, receiving follow-on forces, conducting follow-on operations, sustaining the operations, and conducting decisive operations. Additional activities to consider include—

- Information operations. The full impact of information operation on friendly, neutral, and hostile forces should be considered with the key goal of information operations achieving and maintaining information superiority for the U.S. and its allies; and exploiting enemy information vulnerabilities. Information operations are the integrated employment, during military operations, of information-related capabilities in concert with other lines of operation to influence, disrupt, corrupt, or usurp the decision making of adversaries and potential adversaries while protecting the entry force. (Refer to JP 3-13.)
- Operations security and military deception. Use to confuse the enemy and ease access. Actions, themes, and messages portrayed by all friendly forces must be consistent if military deception is to be believable. Operations security helps foster a credible military deception. (Refer to JP 3-13.3 and JP 3-13.4.)
- Special operations forces. May precede forcible entry forces to include the following:
  - Identify, clarify, and modify conditions in the lodgment.
  - Conduct the assaults to seize small, initial lodgments such as airfields or seaports.
  - Provide or assist in employing fire support.
  - Conduct other operations: seizure airfields, reconnaissance of landing zones or amphibious landing sites.
  - Conduct special reconnaissance and direct action well beyond the lodgment to identify, interdict, and destroy forces that threaten the conventional entry force (See JP 3-05).

1-89. Entry forces must collect, process, and disseminate relevant information in near real time to support fire and movement, and maintain the ability to deliver and control joint fires throughout the assault. Given the distances from which entry forces are deployed and employed, ground forces require access to and direction of joint fires during the assault, stabilization of the lodgment, and introduction of follow-on forces.

1-90. Targeting intelligence supports forcible entry operations in verifying existing information and making recommendations on targeting and collateral damage estimate with respect to intelligence preparation of the operational environment, service capabilities, and rules of engagement. Targeting intelligence analysis encompasses many processes, all linked and logically guided by the joint targeting cycle, that continuously seek to analyze, identify, develop, validate, assess, and prioritize targets for engagement in order to achieve the commander's objectives and end state. (Refer to JP 3-60 and ATP 3-60.1 for more information.)

1-91. The threat a tactical mission may pose to the civilian populace requires balanced lethal and nonlethal actions during forcible entry operations. Overcautious prevention activities or procedures limit the freedom of action just as unrestrained action can result in provocation tactics by adversaries. (Refer to ATP 3-09.32 for more information.)

1-92. Obtaining accurate combat identification provides the ability to differentiate among friendly, enemy, neutral and unknown personnel and objects. *Combat identification* is the process of attaining an accurate characterization of detected objects in the operational environment to support an engagement decision (JP 3-09).

1-93. To achieve surprise or strike a decisive blow, entry forces focus on identifying and defeating enemy area denial capabilities (mine development, emplacement and control network). Commanders must weigh the benefit of massed fires versus the requirement for precision in an area denial environment.

1-94. Contingency response groups planners are integrated into the planning process as early as possible. Contingency response groups provide the USAF first-responder airfield opening unit capability to the JFC regardless of service or mission of the field being opened. Each contingency response group can be scaled to meet specific tasking requirements to support air operations at specific points up to large-scale airfields. (Refer to FM 3-17.2 for more information.)

1-95. Sustainment requirements can be formidable, but must not become such an overriding concern that the forcible entry operation itself is jeopardized. Commanders and staffs must carefully balance the introduction of sustainment forces needed to support initial combat with combat forces required to establish, maintain, and protect the lodgment as well as forces required to transition to follow-on operations.

JOINT ENTRY FORCE

1-96. Forcible entry is executed as either a major operation or a part of a larger campaign to seize and hold a military lodgment in the face of armed opposition for the continuous landing of forces. Joint entry forces can strike directly at the enemy and can open new avenues for other military operations. The entry force employs distributed, yet coherent, forcible entry operations to attack the objective area or areas. The net result is a coordinated attack that overwhelms the adversary before the adversary has time to react. A well-positioned and networked force enables the defeat of adversary reaction and facilitates follow-on operations, if required.

INTELLIGENCE, SURVEILLANCE, AND RECONNAISSANCE

1-97. Joint entry forces must operate a fully integrated and collaborative intelligence, surveillance, and reconnaissance enterprise that provides timely intelligence and counterintelligence to meet entry force requirements. This enterprise must be sustainable and remain responsive and adaptive, and capable of addressing new challenges and opportunities as they emerge. (Refer to JP 2-01 for more information.)

1-98. Entry forces have to fight for and collect information in close contact with the enemy and civilian populations through continuous physical reconnaissance, persistent surveillance, security operations, and intelligence operations: to develop the contextual understanding to counter enemy anti-access capabilities and gain access to an operational area; counter area denial strategies to ensure freedom of action; and adapt continuously to changing situations. (Refer to FM 3-55 and ATP 2-01 for more information.)

1-99. Human interaction on the ground must complement other intelligence to create contextual understanding of events on the ground. This aids in the ability to locate, target, and suppress or neutralize hostile anti-access and area denial capabilities in complex terrain with lethal or nonlethal effects while limiting collateral damage. (Refer to FM 2-22.3 for more information.)

1-100. Long-range surveillance and special operations forces support entry forces by conducting shaping operations in support of forcible entry operations, to include reconnaissance and surveillance and raids. Advanced force operations include strike operations, clandestine insertion of special operations forces and organic reconnaissance teams, deception, counter mine or counter obstacle operations, and information operations. (Refer to FM 3-55.93 and ATTP 3-18.04 for more information.)

Overview

INTEGRATION AND SYNCHRONIZATION

1-101. The complex nature and operational tempo of forcible entry operations requires entry forces to support an increased level of lateral coordination and integration throughout the application of unified action within an operational area. The entry forces' role in the operation for access is the integration of ground and littoral maneuver into the joint effort; conducting entry operations; and employing and sustaining forces while fighting to maintain access throughout the duration of operations.

1-102. Integrating and synchronizing near simultaneous execution of a broad range of operations is essential to presenting the enemy with the greatest range of challenges against which to react. Designing, planning, and executing these operations require a philosophy that embraces decentralization of resources and authority for portions of the force yet more centralized planning and execution for missions where resources are scarce, or strategic sensitivities demand greater control. However, entry forces must be prepared to operate with decentralized decisionmaking, as adversaries may attempt to isolate units by attacking communications capabilities.

REHEARSALS

1-103. Forcible entry is a deliberate operation in that the situation allows for the development and coordination of a specified task organization and a detailed plan, to include multiple branches and sequels. The process of learning, understanding, and practicing a plan in the time available before actual execution reduces and mitigates operational frictions inherent to entry operations.

1-104. Rehearsing key entry force actions and sustainment activities allows participants to become familiar with the operation and the visualization of the plan. This process assists in orienting joint and multinational forces to their surroundings and to other units during execution. Rehearsals provide a forum for subordinate leaders to analyze the plan, but they must exercise caution in adjusting the plan. Changes must be coordinated throughout the chain of command to prevent errors in integration and synchronization.

1-105. While the joint entry force may not be able to rehearse an entire operation, commanders should identify essential elements for rehearsal. Operation plan rehearsal benefits include:
- Common understanding.
- Unity of effort.
- Articulate supporting intents.
- Subordinate and supporting commanders questions.
- Branches or sequels.
- Integration and synchronization.

OPERATIONAL PHASING

1-106. As stated earlier in this chapter, forcible entry operations are conducted during the "Seize the Initiative" or "Dominate" phase of a joint operation. Within the context of these phases established by a higher-level JFC, the joint entry force commander may establish additional phases that fit the forcible entry CONOPS. Planning for each phase must include branch and sequel planning. Transitions between these phases are designed to be distinct shifts emphasized by the joint entry force, often accompanied by changes in command or support relationships. Forcible entry operations may be planned and executed in the following five phases.

Phase I — Preparation and Deployment

1-107. Forcible entry operations are conducted by organizations whose force structures permit rapid deployment into the objective area. Joint entry forces may deploy directly to the operational area or to staging areas to prepare for subsequent operations. Key activities include:
- Planning. All phases, includes Department of Defense agencies and interagency participants.
- Movement. Planning from both strategic and operational perspectives.
- Intelligence. Focused on answering the commander's critical information requirement.
- Reconnaissance and surveillance. Insertion into operational area.

Chapter 1

- Transition to assault. Sets conditions required for successful assault.

Phase II — Assault

1-108. Assault phase, in airborne and air assault operations, a phase beginning with delivery by air of the *assault echelon*— **the element of a force that is scheduled for initial assault on the objective area,** and extending through attack of assault objectives and consolidation of the initial airhead or lodgment area. Key activities include:
- Initial assault designed to surprise and overwhelm the enemy with decisive force and to protect assault force.
- Overcoming natural and man-made obstacles intended to restrict or halt movement that allows the enemy to mass its forces and repel the assault.
- Main assault entry by parachute assault and air assault, landed forces must have immediately available joint fire support.
- Transition to stabilizing the lodgment, introduction of follow-on forces to assist in securing and preparing or repairing the lodgment to allow the landing of air assets, and continue to follow-on operations without an operational pause.

Phase III – Stabilization of the Lodgment

1-109. Stabilization of the lodgment involves securing the lodgment to protect the force and ensure the continuous landing of personnel and materiel.
- Organizing the lodgment to support the increasing flow of forces and logistic resource requirements.
- Expanding the lodgment as required, support the joint force in preparing for and executing follow-on operations.
- Transition to introducing follow-on forces intended to conduct follow-on operations, in extreme circumstances, follow-on forces may be required to assist assault forces in the seizure of initial objectives, or may be used to help secure and defend the lodgment.

Phase IV – Introduction of Follow-On Forces

1-110. The introduction of follow-on forces is required when subsequent operations are planned for conduct in or from the lodgment. It provides the joint force commander with increased flexibility to conduct operations as required.

Phase V – Termination or Transition

1-111. Forcible entry operation to subsequent operations or termination must be an integral part of the planning phase of the joint deployment process. Completed in one of two ways: attainment of the campaign objectives (termination), or completion of the operational objectives when a lodgment is established for follow-on combat operations (transition). (Refer to JP 3-18 for more information.)

SECTION II – VERTICAL ENVELOPMENT

1-112. Operational experience has demonstrated that the rapid projection of combat power is the key to successful ground and littoral maneuver. Experience and analysis have shown that the most effective method of doing so is through a combination of vertical and surface means. These complementary means provide flexibility in negating threats unique to operational access. This section addresses vertical maneuver within forcible entry operations, the airborne or air assault force, and command responsibilities.

TACTICAL APPLICATIONS

1-113. Forcible entry, composed of an entry force, together with other forces that are trained, organized, and equipped for entry operations, project power. Forcible entry can be executed through vertical envelopment, directly against the enemy in a sudden attack in force to achieve decisive results or to establish a lodgment to allow for the introduction of follow-on forces. Besides serving as a forcible entry assault force, such forces can conduct follow-on operations from the lodgment. The following paragraphs discuss the application of vertical envelopment as conducted by airborne or air assault forces to achieve operational objectives.

TACTICAL MANEUVER

1-114. *Vertical envelopment* is a tactical maneuver in which troops that are air-dropped, air-landed, or inserted via air assault, attack the rear and flanks of a force, in effect cutting off or encircling the force (JP 3-18). Commanders conduct vertical envelopment to occupy advantageous ground to shape the operational area and accelerate the momentum of the engagement. An enemy may or may not be in a position to oppose the maneuver. While the commander should attempt to achieve an unopposed landing when conducting vertical envelopment, the assault force must prepare for the presence of opposition.

1-115. Vertical envelopment, airborne and air assault operations, allows a tactical commander to do the following:
- Threaten enemy echelon support areas, causing the enemy to divert combat elements to protect vital bases or installations and hold key terrain.
- Overcome distances quickly, overfly barriers, and bypass enemy defenses.
- Extend the area over which the commander can exert influence.
- Disperse reserve forces widely for survivability reasons while maintaining their capability for effective and rapid response.
- Exploit combat power by increasing tactical mobility.

COMMON FACTORS

1-116. Planning results in establishing positions that support completing the assigned mission. Factors that are common to vertical envelopments include reverse planning process, condition setting, and the impact of meteorological conditions (weather and light data).

Reverse Planning Process

1-117. Airborne and air assault commanders begin planning operations with a visualization of the ground tactical plan and work through a reverse-planning sequence. Planning factors common to airborne and air assault operations are as follows:
- Ground tactical plan. A ground tactical plan is the basis for planning throughout the planning process. It is the first plan completed and it addresses the destruction of enemy forces that pose an immediate threat to the lodgment area. However, each plan affects the others, and changes in one plan can require adjustments in the others.

Note. For example, the amount of lift available determines the feasibility of the ground tactical plan. If there are not enough lift systems to put all the required forces in place at the required time, the commander adjusts the ground tactical plan as well as the other plans. Therefore, vertical envelopment planning requires the unit staff obtain vital planning data, such as the availability of lift systems and the technical and tactical capabilities of those systems, as early as possible.

- Landing plan.
- Movement plan.
- Loading and staging plans. (Air assault.)
- Marshalling plan. (Airborne.)
- Additional planning factors include—
  - In analyzing the plan the commander and staff consider lodgment terrain and infrastructure, with a emphasis on the ability to support follow-on operations and forces.
  - Intelligence regarding the enemy and terrain characteristics of the objective area is vital to this planning process. (Refer to ADRP 2-0 and FM 2-0 for more information.)

- Positive target identification vetted to ensure correct identification. (Refer to ATP 3-09.32 for more information.)
- Continuous coordination between the parallel echelons of the assault force and the supporting forces; from the beginning of an operation until its completion or abandonment.
- Maximized use of combined arms capabilities to ensure the assault force has sufficient power to accomplish its mission and defend itself.
- Short planning times often require staffs to modify contingency plans and SOPs to meet the exact situation while still ensuring adequate coordination.

1-118. The commander determines if adjustments to any of these plans entail acceptable risk. If the risk is unacceptable, the CONOPS changes.

Condition Setting

1-119. Condition setting is an iterative process where the commander's situational understanding determines what part of the situation must change to ensure the success of the vertical envelopment. Warfighting capabilities or functions continually assess until the commander is satisfied with the result or operational necessity forces him to either cancel or conduct the vertical envelopment. Conditions common to airborne and air assault operations are as follows:

- Posture the air assault or parachute force for success with the degree of acceptable risk.
- Suppression of enemy air defense, plan preparation and deception fires.
- Cross-service and echelon-staff, cell, and board synchronization and integration.
- Exchange liaisons.
- Reconnaissance and surveillance detect systems that unacceptably endanger the operational success.
- Service and joint fires-detected targets.

Meteorological Conditions (weather and light data)

1-120. Meteorological conditions influence the conduct of operations. Conditions common to airborne and air assault operations are as follows:

- Impacts vertical envelopment to a greater extent than other operations.
- Long-range forecasts affect planning for force build up and sustainment by aerial delivery.
- Current and future forecasted impact on tactical operations and aircraft performance.
- Current weather information at departure sites and pickup zones, along approach routes, and in the objective area.
- Marginal weather conditions may enhance the element of surprise, but they increase the risk of accidents.
- Deteriorating weather condition: postponement of planned operation or reduced tempo of an ongoing operation when risk becomes unacceptable.
- Weather condition affects on joint fires and medical evacuation or withdrawal of forces.
- Conditions that include wind shears, crosswinds, and the ambient temperatures throughout the course of the operation.
- High temperature and altitude degrade aircraft lift performance— a combination of these factors results in trade-offs in the operating parameters. (For example, a commander may insert dismounted reconnaissance teams on mountainsides in the cool of the morning, but be unable to execute the same mission in the noonday heat.)

AIRBORNE ASSAULT FORCE

1-121. Airborne forces may be used as the assault force or used in combination with other capabilities for a forcible entry; or they may conduct follow-on operations from a lodgment. As an assault force, airborne forces may air land or parachute into the objective area to attack and eliminate armed resistance and secure designated objectives. Airborne forces may be employed from a lodgment in additional joint combat operations appropriate to their training and equipment. Airborne forces offer the JFC an immediate forcible entry option since they can be launched directly from the continental United States and/or forward deployed location without the delays associated with acquiring intermediate staging bases or repositioning of sea-based forces. (Refer to this publication, Part I, Airborne Operations, for more information.)

AIR ASSAULT FORCE

1-122. An *air assault force* is a force composed primarily of ground and rotary-wing air units organized, equipped, and trained for air assault operations (JP 3-18). Air assault forces can deploy from land-based facilities and ships. Fires from aircraft (manned and unmanned) or ships (surface and subsurface) take on added importance to compensate for the lack of artillery. An air assault force may require the establishment of an intermediate staging base. These forces can rapidly project combat power throughout the depth of an operational area. (Refer to Part II, Air Assault Operations, of this publication for more information.)

COMMAND RESPONSIBILITY

1-123. Entry forces conduct airborne and air assault operations with a preference for decentralized decisionmaking. The ability to integrate and synchronize near simultaneous execution of a broad range of operations is essential to presenting the enemy with the greatest range of challenges against which to react. Designing, planning and executing these operations requires a philosophy that embraces decentralization of resources and authority for portions of the force yet more centralized planning and execution for missions where resources are scarce or strategic sensitivities demand greater control.

COMMAND RESPONSIBILITY OF AIRBORNE OPERATIONS

1-124. The airborne assault of a forcible entry operation is delivered by strategic airlift from the continental United States or by strategic or tactical airlift from an intermediate staging base. An airborne assault over intercontinental distance, places additional requirements on joint command and control. Effective employment of an airborne assault force (ABNAF) requires an organizational structure with an Army intermediate higher headquarters nested within the joint task force. This intermediate headquarters, tailored to accompany the assault force to the objective area facilitates the assault force in the execution of its ground tactical plan by controlling enabling functions and units. (See table 1-1, page 1-23.)

Chapter 1

Table 1-1. Command responsibility of airborne operations

JTF CDR / HQ	AIRBORNE TF CDR/HQ (Division)	AIRBORNE ASSAULT FORCE CDR/HQ (IBCT)
Operational planning.	Operational/tactical planning.	Ground tactical planning.
JIPTL development.	Joint fires integration.	Joint fires execution.
Joint ISR integration.	LRS/SOF integration.	Update IPB.
Joint operational access. Forcible entry condition setting.	C2/mission command pre-assault fires.	Receipt of assault fires handover. En route mission planning and rehearsal.
Operational level command and control through JOC, CAOC.	APOE SA and integration en route mission command, airspace control aerial relay using joint communications between JTF/APOE and IBCT/APOD Establish APOD C2/mission command network.	Rig/outload paratroopers and equipment. Conduct airborne assault. Seize assault objective.
Airflow management.	APOE/APOD C2/mission command. RSOI. Generation of combat power Follow-on forces.	Secure APOD. Repair/maintain APOD. Expand lodgment.

LEGEND	
APOD – aerial port of debarkation APOE – aerial port of embarkation APOE SA – aerial port of embarkation staging area CAOC – combat air operations center C2 – command and control CDR – commander HQ – headquarters IBCT – Infantry brigade combat team	IPB– intelligence preparation of the battlefield ISR – intelligence, surveillance, reconnaissance JIPTL – joint integrated priority target list JOC – joint operations center JTF – joint task force LRS – long-range surveillance RSOI – reception, staging, onward movement, and integration SOF – special operations forces

COMMAND RELATIONSHIPS FOR AIRBORNE OPERATIONS

1-125. The airborne assault force commander (ABNAFC) is responsible for seizing the airhead. He accompanies the initial assault and focuses on the planning and execution of the ground tactical plan. Although the size and composition of the assault force varies according to the mission, the ABNAF is typically an Airborne (ABN) Infantry brigade combat team (IBCT).

1-126. The airborne task force commander (ABNTFC) is a higher echelon commander. He may be the joint force commander or ground commander, depending upon experience and the scope of the operation. As the assault requires an ABN IBCT, for example, the BCT commander leads the assault force and the parent Army headquarters (normally a division but could be a corps headquarters) provides the ABNTFC. This ensures that the commander fighting the ground tactical plan can give his full attention to the fight in the airhead without having to manage en route follow-on forces and support. An assault command post of the ABNTFC accompanies the initial ABNAF in order to provide a command element in the airhead to facilitate these functions and to act as an interface with the airborne command post.

1-127. The ABNTFC organizes the parachute assault force, strategic airlift force, supporting fires force, and follow-on airland forces in such a way as to best accomplish the mission. The ABNTFC's responsibilities end upon achievement of a secure airhead line (Refer to chapter 4 of this publication for more information.) and the establishment of either a JFLCC or designated ground commander command post in the airhead.

1-128. En route mission planning and rehearsal systems allow the airborne force to maintain situational awareness and to receive and disseminate updated intelligence while en route from load time until arrival over the airhead.

- Ground commanders in airlift aircraft may communicate with the chain of command over the Army secure en route communications package. Normally, the airlift mission commander and the airborne force commander are in the same aircraft. The senior ground commander can advise embarked ground commanders of changes in the ground tactical situation or to the air movement plan.
- Airborne operations require the use of redundant airborne and ground command posts. Normally, a joint force airborne command post operates from a joint airborne communications center and command post, while a command post from the airborne force operates from a fixed-wing platform with required communications installed.
- TACSAT downlink and other en route communications systems can be used to communicate with USAF special tactics teams, air mobility liaison officers, contingency response elements, and contingency response teams in objective areas.
- The use of special navigational aids and homing devices to direct aircraft to specified areas (for example, a designated drop zone) may be needed. Specialized airborne or air assault force personnel (for example, special tactics teams or long-range surveillance units) are equipped with navigational aids, global positioning systems, and homing devices. These teams are employed early to guide the airborne units, and provide reconnaissance, surveillance, visual flight rules service, and limited instrument flight rules air traffic control service. Other joint force assets such as special operations forces are capable of performing some of these functions.

COMMAND RESPONSIBILITY OF AIR ASSAULT OPERATIONS

1-129. Air assault operations embody the combined arms concept through coordination and planning between the air and ground commanders. Infantry and air units are fully integrated with other members of the combined arms team to form a powerful and flexible joint task force. An air assault operation dramatically extends the commander's ability to influence operations within the area of operations and to execute operations in locations beyond the capability of more conventional forces. Effective employment of an air assault force requires an organizational structure with intermediate higher headquarters nested within the joint task force. (See table 1-2, page 1-25.)

Chapter 1

Table 1-2. Command responsibility of air assault operations

	JTF CDR / HQ	AIR ASSAULT TF CDR / HQ (Div/BCT/BN)	AIR MISSION CDR / HQ (BDE/BN/CO)	GROUND TACTICAL CDR / HQ (BCT/BN/CO)
Planning	Operational planning.	Operational/tactical planning.	Air movement and landing planning.	Ground tactical, loading, and staging planning
Execution	Operational C2 through JOC, CAOC.	Air space management PZ/LZ selection and PZ control. JFE condition monitoring. Aerial relay using joint comms between JTF/PZ and BCT/LZ.	Air movement of personnel and equipment. OPCON of all aviation elements.	Conduct air assault. Seize assault OBJs. Secure LZ. Expand lodgment.
Mission Command	Airflow management.	PZ and follow-on forces control. Go/No Go criteria assessment and decisions.	Mission command from PZ to LZ (AATFC colocated with AMC).	Mission command of all elements cleared from LZ.
Fires	Joint Integrated Prioritized Target List (JIPTL) development.	Joint fires planning/integration. Pre-assault fires/Joint SEAD/SEAD. Clear and direct all fires outside the airhead line.	En route fires. Facilitation from PZ to LZ.	Clear and direct all fires inside the airhead line.
Decision Authority	Personnel recovery operations authority (above immediate unit level).	Approve, disapprove, or modify all components of the assault plan.		
Legend:				

AMB - air mission brief
AMC - air mission commander
AATFC – air assault task force commander
BCT – brigade combat team
CAOC – combat air operations center
C2 – command and control
CDR – commander
HQ – headquarters
JFE – joint fires element

JIPTL – joint integrated prioritized target list
JOC – joint operations center
JTF – joint task force
LZ – landing zone
OBJ – objective
OPCON – operational control
PZ – pickup zone
SEAD – suppression of enemy air defenses

COMMAND RELATIONSHIPS FOR AIR ASSAULT OPERATIONS

1-130. The AATF is a combined arms force under the command of a single headquarters consisting of Infantry, assault and attack reconnaissance helicopters, fire support, electronic warfare, and sustainment assets. The commander ordering the air assault designates the AATF commander. The AATF commander's headquarters coordinates airspace with other airspace users, to include artillery, air defense, unmanned aircraft, close air support, and other aviation units. It coordinates the AATF's plans for maneuver and sustainment with those of higher, subordinate, and adjacent units.

1-131. Normally, a BCT commander serves as the AATF commander for a ground maneuver battalion-size air assault and a ground maneuver battalion commander serves as the AATF commander for a company-size air assault. The air mission commander and the ground maneuver unit commander are subordinate to the AATF commander.

1-132. The air mission commander is the aviation unit commander or his designated representative. The air mission commander receives and executes the guidance and directives from the AATFC, and controls all aviation elements. The air mission commander ensures continuity of command for all supporting aviation units and employs attack helicopters and artillery along the air route, fighting the battle from the pickup zone to the landing zone, while keeping the AATFC informed. The air mission commander has operational control of assault helicopters providing lift to the ground maneuver force and the aviation unit providing assault helicopters is either in direct support of the ground combat unit or under the operational control of the AATF. The support relationship may end at a predetermined point during the operation, on order of the higher commander, or the AATF commander may determine with the air mission commander's input when the operational control relationship begins and ends.

1-133. The commander directing the air assault normally does not attach aviation airlift or escort units to the AATF, because it is unlikely that a ground unit can control the aviation unit and supply the aviation-specific munitions and large amounts of fuel required by aviation units. Direct support and the operational control command relationships do not place logistics responsibility for the supporting unit on the supported unit. The operational control command relationship allows the AATF commander to reorganize the aviation airlift and escorting units when necessary as dictated by the situation. The direct support relationship allows the overall commander to shift the support of these aviation units to other units in response to unexpected developments. Consequently, direct support or operational control is usually the desired relationship between air and ground units in air assault operations.

1-134. The commander ordering the air assault considers the availability and allocation of assault and attack aviation assets when determining the AATF's task organization. He ensures that the ground maneuver force contains sufficient combat power to seize its initial objectives and defend its landing zones. The ground maneuver force requires a mission specific balance of mobility, combat power, and sustainment capabilities. The available rotary-wing aircraft must be able to insert the required combat power into the objective area as quickly as possible to provide surprise and shock effect, consistent with aircraft and pickup zone and landing zone capabilities.

1-135. Air assault operations require the use of redundant airborne and ground command posts. Normally, a joint force airborne command post operates from a joint airborne communications center and command post, while a command post from the air assault force operates from fixed-wing or rotary-wing platform with required communications installed or a specially configured mission command rotary-wing aircraft.

SPACE OPERATION

1-136. Airborne and air assault forces rely on space-based capabilities and systems for precision, navigation and timing, communication, terrestrial and space weather, and intelligence collection platforms to be successful during forcible entry operations. (Refer to JP 3-14 and FM 3-14 for more information.) These systems are critical enablers to plan, communicate, navigate and maneuver, maintain situational awareness, engage the enemy, provide missile warning, and protect and sustain the entry force.

1-137. Space specialists supporting tactical planning by providing expertise and advice regarding available space capabilities and limitations enable space operations. Planning and coordination of space support with national, service, joint, and theater resources takes place with Army space professionals who are attached at the corps and division levels to provide expertise and advice to the commander on space related issues that may impact operations.

1-138. Space support capabilities include receiving accurate status of positioning, navigation, and timing for planning operations, providing capabilities and limitations of space-based intelligence, surveillance and reconnaissance, weather, and communication systems, as well as providing assistance and notification of deliberate enemy interference activities such as attempts to jam or spoof friendly communications.

Chapter 1

1-139. Space-based systems enable airborne and air assault forces and subsequent follow-on forces by—
- Providing rapid communications that enable a commander to gain and maintain the initiative by developing the situation faster than the enemy can react.
- Maintaining a shared common operational picture.
- Retaining the ability to recognize and protect own and friendly forces, as well as synchronize force actions with adjacent and supporting units.
- Providing communication links between forces and commanders within theater and worldwide.
- Monitoring terrestrial areas of interest through information collection assets to help reveal the enemy's location and disposition, and attempting to identify the enemy's intent.
- Providing global positioning system status and accuracy of positioning, navigation, and timing for planning and conducting mission operations such as support for targeting.
- Providing update of solar environment and the impact to both terrestrial and space-based segments of friendly communication systems.
- Providing meteorological, oceanographic, and space environmental information which is processed and analyzed to produce timely and accurate weather effects on operations.

PART I

Airborne Operations

Chapter 2

Organization and Employment

An *airborne operation* involves the air movement into an objective area of combat forces and their logistic support for execution of a tactical, operational, or strategic mission (JP 3-18). The means employed may be any combination of airborne units, air transportable units, and types of transport aircraft, depending on the mission and the overall situation. This chapter focuses on the role, organization, and capabilities of the airborne assault force (ABNAF) as well as the duties and responsibilities of personnel within or task-organized to an airborne (ABN) Infantry brigade combat team (IBCT) for airborne operations.

SECTION I – AIRBORNE ASSAULT FORCE

2-1. An *airborne assault* is the use of airborne forces to parachute into an objective area to attack and eliminate armed resistance and secure designated objectives (JP 3-18). An ABNAF comprises an ABN IBCT with capabilities organized, trained, and equipped to gain entry into an operational area to enable the conduct of follow-on operations or conduct a singular mission.

2-2. The ABNAF seizes an airhead to destroy or capture enemy forces; repel enemy assaults by fire, close combat, or counterattack; for follow-on forces; or for any combination. It can deploy rapidly and be sustained by an austere support structure; and can conduct operations against conventional and unconventional enemy forces in all types of terrain and climate conditions. This section addresses how an ABNAF is organized to include distinct levels of echelon employment and application to conduct an airborne assault

ORGANIZATION OF FORCES

2-3. Once the commander determines the principal features of the ground assault plan (scheme of maneuver and fire support), he task-organizes subordinate units to execute assigned missions and determines boundaries. To ensure unity of effort or to increase readiness for combat, part or all of the subordinate units of a command can be formed into one or more temporary tactical groupings (teams or task forces), each under a designated commander. No standard organization can be prescribed in advance to meet all conditions. Infantry units usually form the tactical nucleus of the team; Infantry unit commanders lead the teams. These teams are tailored for the initial assault by the attachment of supporting units. These supporting units join the Infantry units as soon as possible in the marshalling area to plan and prepare for the initial assault. Dependent on the ground tactical plan, certain supporting units may be detached once centralized control is gained. Other units such as higher echelon command posts can be attached for the movement only.

Chapter 2

INFANTRY BRIGADE COMBAT TEAM

2-4. As the assault requires an ABN IBCT, the BCT commander leads the assault force and the parent Army headquarters (normally a division but could be a corps headquarters) provides the ABNTFC. This ensures that the commander fighting the ground tactical plan can give his full attention to the fight in the airhead and not be consumed with managing en route follow-on forces and support. Although the size and composition of the ABNAF varies according to the mission, the ABNAF is typically an ABN IBCT that is scalable and tailored with additional capabilities and forces as determined by the ABNTFC.

INFANTRY BATTALION

2-5. For control, the airborne Infantry battalion usually is reinforced for the airborne assault and is organized into a task force. This is especially true if battalions land in widely separated drop zones or landing zones. A battalion task force usually comprises an airborne Infantry battalion with reinforcements based on the IBCT commander's estimate for the airborne assault. *Follow-on echelon— those additional forces moved into the objective area after the assault echelon.* Follow-on echelon reinforcements may include more Infantry, Armored, Stryker, cavalry, antitank, engineer, dedicated artillery, and other units or detachments needed to expand the lodgment. As in the IBCT, attachments to the task force for the airborne assault are made early in the planning phase. They can be withdrawn as soon as the ground situation stabilizes.

INFANTRY RIFLE UNITS

2-6. Airborne Infantry rifle companies and platoons can be reinforced for the airborne assault according to the usual considerations governing a ground attack. Attachments are made before the move to, or on arrival in, the marshalling base.

CAVALRY SQUADRON

2-7. The cavalry squadron of the airborne IBCT, assigned to perform reconnaissance and security missions within the assault phase, usually is reinforced for the airborne assault into a task force. After the ABNAF makes the initial assault landing into the objective area and consolidates the initial airhead, the commander organizes the airhead line, confirms reconnaissance and security plans, and task-organizes the cavalry squadron for operations in the security area.

ECHELON EMPLOYMENT

2-8. After the task organization of units or Soldiers for the airborne assault or landing is announced, units organize into assault, follow-on, and rear echelons. The airborne IBCT as the ABNAF, employs organic forces and other attached units assigned for the mission over three echelons: the assault echelon, the follow-on echelon, and the rear echelon. Elements of the higher headquarters ABNAF are employed throughout the three echelons as directed by the ABNAFC.

ASSAULT ECHELON

2-9. The assault echelon (airborne assault) referred to as the Alpha Echelon, is the initial entry force. It is part of the ABNAF that conducts the parachute assault on an unsecured drop zone to seize the lodgment or initial assault objectives. The airborne assault echelon is composed of those forces required to conduct the parachute assault to seize assault objectives and establish the initial airhead, and if appropriate, prepare an airfield to receive follow-on echelons. This echelon is deployed with sufficient supplies to sustain operations for 72 hours. It includes the assault command post of the ABNTF headquarters providing joint interoperability.

FOLLOW-ON ECHELON

2-10. The follow-on echelon comprises two elements. Referred to as the Bravo Echelon and the Charlie Echelon, these elements are the airland portion of the ABNAF.

- The Bravo Echelon is the reinforcing forces in the airhead. When needed, the Bravo Echelon enters the objective area as soon as practical by air or surface movement, or a combination of the two. It is manned and equipped for combat power augmentation and lodgment expansion. It includes additional vehicles and equipment from the ABNAF, plus more forces to include supporting personnel. The existence of one of the following conditions requires an ABNAF to have a follow-on echelon:
    - Shortage of aircraft.
    - Aircraft that cannot land heavy items of equipment.
    - Any enemy situation, terrain, or weather that makes it impossible to land certain Soldiers or equipment in the assault echelon.
- The Charlie Echelon includes the remainder of the ABNAF and the airborne task force (ABNTF) required for operations after the establishment of the airhead and securing or expanding the lodgment. The Charlie Echelon gives the ABNAF the capability to conduct sustained combat operations within the limitations of the IBCT and provides additional combat power and equipment for the ABNTF. Depending on the mission and subsequent operations, the Charlie Echelon may be either a reinforcing force or the first elements of what will become the follow-on force.

Note. Airborne Infantry units can be committed to an airborne assault with the full complement of the follow-on echelon as part of the assaulting force; however, if leaving a follow-on echelon that must be brought forward by means other than air, it is often desirable or necessary to leave certain personnel and equipment behind.

REAR ECHELON

2-11. The ***rear echelon* is the echelon containing those elements of the force that are not required in the objective area.** The rear echelon is normally small for a brigade or battalion and includes personnel left at its rear base to perform administrative and service support functions. A higher headquarters usually controls the rear echelon for all units. The rear echelon can remain at the remote marshalling base when the unit is to be relieved at an early date; or it can rejoin the unit when the IBCT remains committed to sustained combat for a prolonged or indefinite period. In addition, if the airborne force continues in the ground combat role after linkup, the rear echelon may be brought forward.

DESIGNATION OF RESERVE

2-12. The employment of the reserve element follows the normal employment of a reserve unit in a ground operation. The location occupied by the echelon reserve depends on the most likely mission for the reserve on commitment. With the reserve element at the departure airfield, the reserve commander must continue planning for possible future commitment of his forces as far as maps, photos, and information of the situation permit.

The Battalion as the Reserve

2-13. The reserve can be held in the departure area ready to be committed by air when and where the situation dictates. This usually happens in large-scale airborne operations when suitable airfields in the airhead are not available. However, it may cause delays in commitment—
- If signal communications fail.
- If the air movement is long.
- If flying weather is unfavorable.
- If time is added for coordination of air cover.

Brigade and Battalion Reserves

2-14. These reserves may be used to enter the airhead as part of the assault echelon. They provide depth to the airhead by blocking penetrations, reinforcing committed units, and counterattacking. They consist of not more than a company at BCT level or a platoon at battalion level. However, their small size is dictated by tactical considerations and assigned missions. Commanders should organize, task, and position the reserve, ensuring that—

- The size of the reserve is compatible with likely missions.
- The reserve comes from the unit with the fewest priority tasks.
- The reserve is not assigned assault objectives or an area of the airhead to defend.
- The reserve is positioned in an area that allows for quick employment.
- The reserve is mobile. (This can be achieved using organic vehicles (such as, weapons company, forward support company, or if augmented with Armored or Stryker forces.)
- The reserve is located in an assembly area, both initial and subsequent assembly areas, or a battle position, so that it does not interfere with units assigned assault objectives.
- The reserve is near lines of communication in a covered and concealed location to provide ease of movement, to reinforce, or to block.
- The reserve is located within the area of one unit, if possible.
- The reserve's location allows for dispersion of the force.

SECTION II – AIRBORNE ASSAULT FORCE CONSIDERATIONS

2-15. The ABNAF can be part of a larger unit, or it can be comprised solely of the initial assault force, preparing the way for deployment of a follow-on force. This section discusses assault force considerations for formation and support.

ASSAULT FORCE FORMATION

2-16. The ABNAF is formed early in the planning stage by a directing or establishing headquarters that allocates units and defines authority and responsibility by designating command and support relationships. Predesignated and well-understood command and support relationships include—

- A task organization that provides a mission-specific balance of maneuver, combat power, and endurance to seize and protect the airhead or lodgment.
- Unit tactical integrity that is maintained when developing load plans and cross loading key leadership, crew-served weapons and equipment followed by all other personnel to ensure unit integrity upon insertion into the drop zone and assemble.
- A sustainment capability to support a rapid tempo until follow-on or linkup forces arrive, or until the mission is completed.

JOINT AND ARMY CONTROL TEAMS

2-17. Terminal guidance aids and control measures are used on the ground in the objective area to assist and guide incoming airlift aircraft to the designated drop zones or landing zones. Combat control teams comprised of USAF personnel are organized, trained, and equipped to provide aircraft terminal guidance. Army teams from the long-range surveillance company (LRSC), a divisional or corps asset, are organized, trained, and equipped to deploy into the objective area and conduct reconnaissance and surveillance tasks before the deployment of the airborne force.

COMBAT CONTROL TEAM

2-18. The combat control team is a small task-organized team of USAF parachute and combat diver qualified personnel. The combat control team's mission is to establish assault zones (drop zones and landing zones) in austere and non-permissive environments. The mission includes initially placing en route and terminal navigational aids; controlling air traffic; providing communications; and removing obstacles and unexploded ordnance with demolitions. Combat control teams provide command and control, reconnaissance and surveillance, and limited weather observations. (Refer to JP 3-17 for more information.)

Organization and Employment

LONG-RANGE SURVEILLANCE COMPANY

2-19. The long-range surveillance company (LRSC) is a corps level asset for the purpose of long-range surveillance. The LRSC comprises a headquarters section, a communications platoon, a transportation section, a maintenance section, and three long-range surveillance platoons with three teams each for a total of 9 teams. The LRSC is modular in that it has the command, control, and communications capability to support multiple operations simultaneously. (Refer to FM 3-55.93 for more information.)

SECTION III – CAPABILITIES, LIMITATIONS, VULNERABILITIES

2-20. Airborne forces deploy strategically, operationally, or tactically on short notice anywhere in the world. They can be employed as a deterrent or as a combat force. The strategic mobility of airborne forces permits rapid employment to meet contingence across the range of military operations; and provides a means by which a commander can decisively influence operations. This section discusses the capabilities, limitations and vulnerabilities unique to an ABNAF.

CAPABILITIES

2-21. Since airborne forces are able to respond on short notice, airborne operations provide distinct advantages such as—
- Ability to bypass all land or sea obstacles.
- Surprise.
- Ability to mass rapidly on critical targets.

2-22. Airborne forces can extend the area of operation, move, and rapidly concentrate combat power like no other available forces. Specifically, airborne forces can—
- Attack enemy positions from any direction.
- Conduct attacks and raids beyond the area of operation.
- Conduct limited exploitation and pursuit operations.
- Overfly and bypass enemy positions, barriers, and obstacles and strike objectives in otherwise inaccessible areas.
- Provide responsive reserves, allowing commanders to commit a larger portion of their forces to action.
- React rapidly to tactical opportunities, necessities, and threats in unassigned areas.
- Rapidly place forces at tactically decisive points in the area of operation.
- Conduct fast-paced operations over extended distances.
- Conduct and support deception with false insertions.
- Rapidly reinforce committed units.
- Rapidly secure and defend key terrain (such as crossing sites, road junctions, and bridges) or key objectives.
- Rapidly repair or construct infrastructure to receive follow-on forces.
- Delay a much larger force without becoming decisively engaged.

2-23. Airborne forces, when augmented with appropriate support or augmentation, can conduct sustained combat operations against the enemy.

LIMITATIONS

2-24. The commander and planners must recognize the limitations of airborne forces and plan accordingly. They must consider the following:
- An airborne force depends on USAF aircraft for long-range movement, fire support, and sustainment. The availability and type of aircraft dictates the scope and duration of airborne operations.
- After the initial airdrop, the sustained combat power of airborne forces depends on resupply by air. Any interruption in the flow of resupply aircraft can cause a potential weakening of the

airborne force. Enemy air defense fires against resupply aircraft and long-range artillery and mortar fires on the drop zone can hamper the delivery, collection, or distribution of critical supplies.

- Once on the ground, the airborne force has limited tactical mobility. That mobility depends on the number and type of vehicles and helicopters that can be brought into the objective area with the follow-on force.
- The airborne force has limited field artillery and air defense artillery support until additional assets can be introduced into the objective area. Additional target acquisition assets are needed to provide accurate and timely targeting information.
- Evacuation of casualties from the airhead is difficult. Until evacuation means are available, the BCT must be prepared to provide medical care through the attachment of its organic medical company or the attachment of echelon above BCT medical elements.

VULNERABILITIES

2-25. Airborne forces are vulnerable to enemy attack while en route to the drop zone. Although the USAF can conduct limited airdrops without air superiority, large operations require neutralization or suppression of enemy air defenses (SEAD). This may require SEAD, radar jamming, and fighter aircraft besides transport and close air support sorties. Initial airborne assault elements are light and are separated from weapon systems, equipment, and materiel that provide protection and survivability. An ABNAF is particularly vulnerable to enemy—

- Attack by aircraft and air defense weapon systems during the movement and airborne assault phases.
- Attack by chemical, biological, radiological, and nuclear weapons because of limited chemical, biological, radiological, and nuclear protection and decontamination capability.
- Attack by ground, air, or artillery during the assault and landing phases.
- Air strikes if air superiority is not gained before the airborne assault.
- Electronic attack, to include jamming of communications and navigation systems, and disrupting aircraft survivability equipment.
- Small-arms fire that presents a large threat to the aircraft during the air movement, airborne assault and landing phases.

SECTION IV – AIRBORNE COMMAND AND CONTROL PLATFORMS

2-26. Airborne operations require extensive coordination between the USAF, Army, and often, other services. During forcible entry, airborne command and control platforms may be employed separately or in combination to augment or even replace the ground-based elements when response time is critical. This section discusses the airborne elements of the tactical air control system, consisting of the Airborne Warning and Control System (AWACS), Joint Surveillance Target Attack Radar System (JSTARS), and the Joint Airborne Communications Center/Command Post (JACC/CP) or "Jackpot" package, designed to fit aboard a C-130 aircraft.

AIRBORNE WARNING AND CONTROL SYSTEM

2-27. The AWACS, designated by the USAF as the E-3 radar, is a modified Boeing 707 that houses a radar subsystem and vast communications equipment. It is under operational control of the tactical airlift control center (TACC). The AWACS radar system can compensate for the major limitations of ground-based radar systems such as their inability to detect low-flying aircraft due to line-of-sight restrictions. Other limitations of ground-based radar systems include their susceptibility to electronic countermeasures and their vulnerability to attack.

2-28. The AWACS can communicate with a wide range of systems. It has extensive high frequency (HF), very high frequency (VHF), and ultrahigh frequency (UHF) radios used to communicate with ground controllers, airborne forces, and ground forces. The E-3's radar flexibility allows it to support tactical missions, defensive missions, or both at the same time. The aircraft is used for weapons control or as a

Organization and Employment

surveillance platform. In an air defense role, the E-3 radar provides weapons control and surveillance capabilities. It provides control for weapons and air defense regions during stages of increased alerts.

JOINT SURVEILANCE TARGET ATTACK RADAR SYSTEM

2-29. The E-8C Joint Surveillance Target Attack Radar System (JSTARS) is an airborne command and control, intelligence, surveillance and reconnaissance platform. Its primary mission is to provide theater ground and air commanders with ground surveillance to support attack operations and targeting that contributes to the delay, disruption and destruction of enemy forces. When available, JSTARS aircraft may be employed as a viable airborne assault force command platform in support of forcible entry operations.

2-30. The E-8C is a modified Boeing 707-300 series commercial airframe extensively remanufactured and modified with the radar, communications, operations and control subsystems required to perform its operational mission. The most prominent external feature is the 27-feet (8 meters) long, canoe-shaped dome under the forward fuselage that houses the 24-feet (7.3 meters) long, side-looking phased array antenna. This aircraft is capable of in-flight refueling, allowing flexibility for extend operations in support of forcible entry operations.

2-31. The radar and computer subsystems on the E-8C can gather and display detailed information on ground forces. The information is relayed in near-real time to ground command, control, communications, computers and intelligence systems. The antenna can be tilted to either side of the aircraft where it can develop a 120-degree field of view covering nearly 19,305 square miles (50,000 square kilometers) and is capable of detecting targets at more than 250 kilometers (more than 820,000 feet). The radar has some limited capability to detect helicopters, rotating antennas and low, slow-moving fixed wing aircraft.

JOINT AIRBORNE COMMUNICATIONS CENTER/COMMAND POST

2-32. The Joint Airborne Communications Center/Command Post (JACC/CP), or "Jackpot" package is designed to fit aboard a military C-130 aircraft and has military and civilian band radios and Internet access. Joint Communications Support Element (JCSE) is the contingency support unit for the package consisting of U.S. Army, Navy, Air Force, and Marine Corps personnel providing operator support to assist G-6 in managing the variety of communications equipment to include: SECRET Internet Protocol Router Network (SIPRNET) services through an international maritime satellite antenna; as well as, a full array of FM and TACSAT networks. The JACC/CP can be deployed within 24 hours from the time the JCS issues deployment approval messages. With most C-130 aircraft not being capable of in-flight refueling, time available over an objective may be limited, possibly requiring basing from a forward area or an intermediate staging base. (Refer to JP 6-0 for more information.)

2-33. The JACC/CP has four major components: operations center (12 SIPRNET laptop workstations), communications control, generator, and an air conditioner/accessory trailer. The jackpot package provides one high frequency, single sideband (HF/SSB) voice or teletype communication channel over its 1-kilowatt transceivers or high frequency, double independent sideband with a total of four independent 3 kilohertz (3 SPKHZ) voice or teletype channels over its 10-kilowatt system. The 10-kilowatt system is limited to ground operations only. The JACC/CP contains three radios, an AN/ARC-73 (VHF/AM), AN/ARC-54 (VHF/FM) and AN/ ARC-51BX (VHF/AM), for ground-to-ground and ground-to-air communications.

2-34. The voice radio system may be connected to a 10-line, 20-line, or 30-line, four-wire/two-wire telephone switchboard. The switchboard can connect any telephone subscriber to another telephone or a JACC/CP. The complete JACC/CP can be transported in a winch equipped C-130 or larger aircraft. A wide lowboy trailer must be used to transport the vans any distance or over other than paved or gravel roads.

This page intentionally left blank.

Chapter 3
Airborne Assault Planning

Airborne assaults may be conducted as a rapid crisis response against less capable enemies, where the conditions needed for the entry are quickly set with limited shaping operations or where forward deployed and rapid response elements must conduct the airborne assault mainly with their organic capabilities and minimal reinforcement. They may be larger-scale entry operations where there is significant shaping required to set the conditions for the airborne assault. Commanders begin planning for an airborne assault with a visualization of the ground tactical plan and work backwards through the landing plan, the air movement plan, and the marshalling plan. Planning is conducted in this order regardless of the type and duration of the mission or the size of the force. This chapter addresses roles and responsibilities, planning sequence and considerations, and shaping operations for an airborne assault.

SECTION I – ROLES AND RESPONSIBILITIES

3-1. Airborne assault planning is as detailed as time permits and should include completion of written orders and plans. Within time constraints, the ABNAFC carefully evaluates capabilities and limitations of the total force and develops a plan that communicates a common vision and synchronizes the action of forces in time, space, and purpose to achieve objectives and accomplish missions. The planning should be highly structured involving the commander, staff, subordinate commanders, and others to develop a fully synchronized plan or order. Planning time should abide by the one-third/two-thirds rule to ensure subordinates have enough time to plan and rehearse.

HIGHER HEADQUARTERS

3-2. The joint task force commander directs the composition of the ABNTF headquarters and the ABNAF. This headquarters allocates units, defines authority, and assigns responsibility by designating command and support relationships. The staff of this headquarters is responsible to develop the task organization of the ABNAF and conducting the necessary steps of the military decisionmaking process (MDMP). A division-level commander or his equivalent is the approving authority for the formation of an ABNAF. (Refer to FM 3-94 for more information.)

PLANNING RESPONSIBILITIES

3-3. The joint task force commander initiates airborne operations with a planning directive to participating units. The directive assimilates through normal command channels at the corps and division levels; pertinent information then is passed to BCTs. The directive must—
- Specify missions.
- Outline the command structure.
- Identify participating ground and air forces.
- List forces in support.
- Provide a schedule of events.
- State conditions under which the operation begins, is delayed, is altered, or is terminated.
- Establish supported and supporting relationships.

Chapter 3

KEY PERSONNEL RESPONSIBILITIES

3-4. The ABNTFC establishes mission command by ensuring that his concept is understood and by defining the responsibilities of key personnel for—
- Accomplishing the ground mission.
- Task-organizing and aircraft assignment.
- Sustaining.

3-5. The commander, airlift force for defining responsibilities of key personnel for—
- Aircraft allocation to support the ground tactical plan.
- Assault force insertion.
- Resupply and evacuation.

3-6. The ABNTFC and commander, airlift for defining joint responsibility of key personnel for—
- Establishing control parties at departure location.
- Loading (Soldiers and equipment).
- Rehearsing.
- Rehearsing communication coordination and standardization.
- Selecting drop zones and landing zones.
- Establishing control parties at drop zones and landing zones.
- Uploading aircraft.
- Planning aerial resupply and evacuation.
- Departing airfield security.
- Working or planning air movement tables.
- Coordinating movements (Soldiers and aircraft).

AIRBORNE ASSAULT FORCE

3-7. The ABN IBCT is the core of the ABNAF and the ABN IBCT commander is normally the ABNAFC for a battalion or larger airborne assault. The primary role of the ABNAFC and his staff is to develop the ground tactical plan by providing his staff and all supporting unit commanders and staffs with key tasks, intent and guidance concerning the weight of the attack, reconnaissance coverage and the level of acceptable risk.

SUPPORTING AVIATION

3-8. Strategic location of the airhead or lodgment may limit what aviation support can be provided for reconnaissance and attack of the airhead or lodgment before the airborne assault. Once the joint force commander relinquishes control of the operation to the ABNAFC, both fixed- and rotary-wing supporting aviation units are under the operational control of the ABNAFC. Not all fixed-wing aviation falls under the ABNAFC, only that which is in direct support of the ground tactical plan.

3-9. Once the airhead or lodgment is established, and dependent on the ground tactical plan, an aviation task force may be created to support an ABNAF especially if the ground tactical plan necessitates support for follow-on operations. However, the combat aviation brigade commander typically anticipates the needs of the ABNAFC and provides the necessary aviation units to support the mission of the ABNAF. As the supporting unit, the combat aviation brigade commander directs aviation units within his command or requests augmentation from his higher headquarters to meet the needs of the ABNAFC.

SECTION II – REVERSE PLANNING SEQUENCE

3-10. The ABNTFC and his staff develop, in this order, the ground tactical plan, the landing plan, the air movement plan, and the marshalling plan. The ABNAF staff and all supporting units coordinate, develop, and refine concurrently to make best use of available time and resources. They develop first the ground tactical plan, which serves as the basis to develop the other plans. Each plan may potentially affect the others. Changes in an aspect of one plan may require adjustments in the other plans. The ABNAFC must determine if such adjustments entail acceptable risk. If the risk is unacceptable, the concept of operations (CONOPS) must change.

GROUND TACTICAL PLAN

3-11. The ground tactical plan is the basis for the development of all other plans. The ABNAFC and his staff give special consideration to the assembly and organization of the assault forces and to the decentralized nature of initial operations in the objective area. The subordinate commander requires the ground tactical plan of his higher headquarters before he can begin planning. He needs to know the type, location, and size of objectives and the enemy situation at each one; the mission and intent of higher headquarters two levels up; and his task and purpose. The ground tactical plan is generated down the chain of command as a mutual effort. (Refer to chapter 4 of this publication for more information.) The ground tactical plan includes—

- Assault objectives and airhead line.
- Reconnaissance and security forces to include observation posts.
- Boundaries.
- Task organization.
- Designation of reserve.
- Supply (accompanying, follow-up, routine).
- Fire support plan.
- Tactical cross load (for air land or parachute).

LANDING PLAN

3-12. The landing plan is the ABNAFC's plan that links the air movement plan to the ground tactical plan. It is published at brigade level and below. Before the ABNAFC can prepare an overall landing plan, he must know where the subordinate commander wants to place his assault force. The landing plan is generated up the chain of command as a mutual effort. (Refer to Chapter 5 of this publication for more information.) The landing plan includes—

- Drop zone, landing zone, locations and descriptions.
- Sequence of delivery.
- Method of delivery.
- Place of delivery.
- Time of delivery.
- Cross-loading plan.
- Assembly plan.
- Landing plan worksheet.

AIR MOVEMENT PLAN

3-13. The air movement plan provides the information required to move the airborne force from the departure airfields to the objective area. This plan is the third step in the reverse planning process and covers the period from when units load to when they exit the aircraft. The airborne commander designates the subordinate unit's sequence of airflow and allocates aircraft. This allows the subordinate commanders to conduct air movement planning. The air movement plan is generated up the chain of command as a mutual effort. (Refer to chapter 6 of this publication for more information.) The air movement plan includes—

- Departure airfields.
- Aircraft by serial.
- Parking diagram.
- Aircraft mission (air movement tables and flight routes).
- Unit providing the aircraft.

MARSHALLING PLAN

3-14. This plan is developed last in the reverse planning sequence and is based on the requirements of the other plans. It provides the needed information for units of the assault force to prepare for combat, to move to departure airfields, and to load aircraft. The marshalling plan provides detailed instructions for facilities and services needed during marshalling. It is generated down the chain of command. (Refer to chapter 7 of this publication for more information.) The marshalling plan includes—

- Movement to the marshalling area.
- Passive defensive measures.
- Dispersal measures.
- Departure airfields.
- Marshalling operations.
- Confirmation brief schedule.
- Preparation for combat (backbrief, inspection, supervision, rehearsal, and rest).
- Communications.

SECTION III – PLANNING CONSIDERATIONS

3-15. Whether done deliberately or rapidly, all planning requires skillful use of available time to optimize planning and preparation throughout the unit. Taking more time to plan often results in greater synchronization; however, any delay in execution risks yielding the initiative with more time to prepare and act to the enemy. When allocating planning time ensure subordinates have enough time to plan and prepare their own actions before execution.

PLANNING METHODOLOGY

3-16. Planning for airborne operations mirrors the MDMP. (See figure 3-1.) It incorporates parallel and collaborative planning actions necessary to provide the additional time and detailed planning required for successful execution of an airborne assault mission.

Airborne Assault Planning

MDMP Steps	Planning Steps	Key Inputs	Key Outputs	Key Attendees
Step 1: Receipt of Mission	WARNORD	• Higher headquarters' plan or order a new mission anticipated by the commander.	• Commander's guidance. • Initial allocation of time.	
Step 2: Mission Analysis		• Higher headquarters' plan or order. • Higher headquarters' knowledge and intelligence products. • Knowledge products from other organizations. • Design concept (if developed).	• Problem statement. • Mission statement. • Initial commander's intent. • Initial planning guidance. • Initial CCIRs and EEFIs. • Updated IPB and running estimate. • Updated assumptions.	
Step 3: Course of Action (COA) Development		• Mission statement. • Initial commander's intent, planning guidance, CCIRs and EEFIs. • Updated IPB and running estimates. • Assumptions.	• COA statements and sketches. • Tentative task organization. • Broad concept of operations. • Revised planning guidance. • Updated assumptions.	
Step 4: COA Analysis (War-game)		• Updated running estimates. • Revised planning guidance. • COA statements and sketches. • Updates assumptions.	• Refined COAs. • Potential decision points. • War-game results. • Initial assessment measures. • Updated assumptions.	
Step 5: COA Comparision	INITIAL PLANNING CONFERENCE	• Updated running estimates. • Refined COAs. • Evaluation criteria. • War-game results. • Updated assumptions.	• Evaluated COAs. • Recommended COAs. • Updated running estimates. • Updated assumptions.	• BCT S-2, S-3 • Co CDRs • Unit S-2, S-3, S-3 Air • ALO • Others as required
Step 6: COA Approval	AIR MISSION COORDINATION MEETING	• Updated running estimates. • Evaluated COAs. • Recommended COA. • Updated assumptions.	• Commander-selected COA and any modifications. • Refined commander's intent, CCIRs and EEFIs. • Updated assumptions.	• BCT S-2, S-3, S-6 • BAO • ALO • LRSC CDR • FSC CDR • FSO • Others as required
Step 7: Orders Production	ORDERS DEVELOPMENT	• Commander-selected COA and any modifications. • Refined commander's intent, CCIRs and EEFIs. • Updated assumptions.	• Approved operation plan or order.	

LEGEND
ALO — Air Liaison Officer
BAO — Brigade Aviation Officer
BCT — Brigade Combat Team
CCIR — Commander's Critical Information Requirements
COA — Course of Action
EEFI — Essential Elements of Friendly Information
FSC CDR — Forward Support Company Commander
FSO — Fire Support Officer
IPB — Intelligence Preparation of the Battlefield
LRSC CDR — Long-Range Surveillance Company Commander
S-2 — Battalion or Brigade Intelligence Staff Officer
S-3 — Battalion or Brigade Operations Staff Officer
S-6 — Battalion or Brigade Signal Staff Officer

Figure 3-1. MDMP and airborne assault planning process

PREDEPLOYMENT PLANNING AND PREPARATION

3-17. Units must plan for and prepare internal deployment standard operating procedures (SOPs) and continually update and rehearse them. These SOPs should include actions that are common to all deployments, to include airland, parachute assault planning, preparation, execution, and assessment. They may include:

- Conduct rapid, short notice deployment.
- Emergency deployment readiness exercise (commonly known as EDRE).
- Prepare personnel for overseas deployment. (Refer to FM 3-35 for more information.)
- Update and review all vehicle load plans.
- Validate and update movement plans with next higher headquarters.
- Update access and recall rosters.
- Review family readiness group rosters and rear detachment responsibilities.
- Ensure special team personnel are identified and trained (movement, chemical, biological, radiological, and nuclear, outload, ammunition handling).

X-HOUR/N-HOUR SEQUENCES FOR DEPLOYMENT

3-18. The X-hour/N-hour sequences for deployment are developed and followed to ensure all reports, actions, and outload processes are accomplished at the proper time during marshalling. They aid in developing air and deployment schedules and are flexible to allow for modifications based on the mission and the unit commander's concept of the operation.

3-19. **X-hour is the unspecified time that commences unit notification for planning and deployment preparation in support of potential contingency operations that do not involve rapid, short notice deployment. X-hour sequence is an extended sequence of events initiated by X-hour that allow a unit to focus on planning for a potential contingency operation, to include preparation for deployment.**

3-20. **N-hour is the time a unit is notified to assemble its personnel and begin the deployment sequence.** The **N-hour sequence starts the reverse planning necessary after notification to have the first assault aircraft en route to the objective area for commencement of the parachute assault in accordance with the order for execution.**

3-21. In anticipation of an order for execution; the ABNAF staff and its key leaders begin preparing or updating an operations plan. The length of X-hour planning varies based on the contingency planning or crisis action planning situation and the specific operations plan. It normally ceases with either the designation of N-hour, or if political or military events warrant, no further action. Deployment planning sequences fall into one of three scenarios:

- Unconstrained X-hour sequence. Used primarily for deliberate planning or crisis-action planning that is not under a time constraint.
- Constrained X-hour sequence. Used for crisis action planning.
- N-hour sequence. May be proceeded by an X-hour sequence.

OPTIMIZE AVAILABLE PLANNING TIME

3-22. Effective execution requires issuing timely plans and orders to subordinates. Timely plans are those issued soon enough to allow subordinates time to plan, issue their orders, and prepare for operations. In time-constrained environments, products contain just enough information for the commander to make a reasoned decision and subordinates to assess the situation quickly and plan, prepare, and execute the necessary actions. Regardless of whether time for planning is constrained or not, to optimize available time and ensure the best possible synchronization, commanders encourage collaborative and parallel planning between their headquarters and higher and lower headquarters.

COLLABORATIVE AND PARALLEL PLANNING

3-23. Both collaborative and parallel planning help optimize available planning time. Collaborative planning is several echelons developing plans and orders together. Commanders, subordinate commanders, and staffs share their understanding of the situation, and participate in course of action development and decisionmaking for development of the higher headquarters plan or order.

3-24. Parallel planning is two or more echelons planning for the same operation through the sequential sharing of information from the higher headquarters before the higher headquarters publishes its operations plan or operation order (OPORD). It requires significant interaction between echelons. During parallel planning, subordinate units do not wait for their higher headquarters to publish an order to begin developing their own plans and orders.

INFORMATION SHARING

3-25. The higher headquarters continuously shares information concerning future operations with subordinate units through warning orders (WARNORDs) and other means. Frequent communication between commanders and staffs and sharing of information (such as intelligence preparation of the battlefield (IPB) products) helps subordinate headquarters plan.

3-26. Generally, the higher the headquarters has more time and staff resources available to plan and explore options. They are sensitive not to overload subordinates with information and planning requirements. Higher headquarters provide subordinates with information and involve them in the development of those plans and concepts that have the highest likelihood of being adopted or fully developed.

3-27. Commanders provide plans and orders down the chain of command. However, for airborne operations, higher headquarters often cannot complete their plans until subordinate units have conducted a backbrief of their plans as a change in one plan impacts other plans.

3-28. Parallel echelons of the airlift and ABNAF units coordinate continuously from the time of the joint planning conference until the operation is executed or cancelled. They exchange liaison officers to act as advisors and coordinators immediately upon receipt of orders to participate in an airborne assault. ABNAF liaison officers must be familiar with all aspects of the airborne assault. They must attend briefings and conferences, and must be provided with adequate transportation and communications assets. Liaison officers normally are exchanged between the ABNAF and—

- Army units supporting the operation from outside the objective area.
- Close air support and airlift elements.
- Linkup forces.
- Special operations forces. (Refer to FM 3-05 for more information.)

3-29. The specific duties of liaison officers include—

- Represent their unit headquarters at the headquarters to which they are detailed.
- Act as advisors to the headquarters on matters pertaining to their own commands.
- Coordinate matters involving dual responsibility.
- Discuss the time, place, personnel required, and material to be covered at coordination meetings and when necessary hold coordination briefings, both at the joint and service level.
- Examine parallel orders to ensure complete agreement of plans and arrangements.
- Assess and plan for the availability and procurement of equipment and facilities required from the higher headquarters.
- Attend all joint conferences, have active knowledge with the agreements reached by the commanders and with the operations plan.
- Prepare joint reports.
- Obtain copies of the marshalling plan and the parking diagram for their units.
- Know the location and capacity of all installations at the airfields and air landing facilities that concern their units.
- Review the plans and arrangements for replacement aircraft if last minute failures occur; prepare to assist the movement of ABNAF from aborting aircraft to reserve aircraft.

Chapter 3

- Brief guides, who are furnished by the ABNAF, on airfield traffic measures and locations of aircraft to be loaded. At dispersed locations, an ABNAF representative is located at the coordination facility to perform this function and to act as individual liaison.

3-30. Commanders exchange liaison officers on a continuous duty status at echelons higher than ABNAF level. At BCT and lower echelons, the S-3 liaison officers, the S-3 Air, or unit air movement officer can perform these duties. For operations of less than ABN IBCT size, commanders exchange liaison officers as needed.

3-31. When the ABNAF is a follow-on force after a special operations force, it requests a liaison before arrival in the operational area. During the planning phase, a special operations force liaison officer is attached to the ABN IBCT along with all communications assets needed for immediate use with special operations forces assets at joint special operations task force (JSOTF) and at the objective area. The signal plan must standardize not only frequencies and call signs, but address visual signals, and day and night operations as well. (Refer to FM 3-05 for more information.)

3-32. Subordinate commanders must conduct confirmation briefs and backbriefs on all aspects of their plan to the next higher commander. The backbrief differs from the confirmation brief (a briefing subordinates give their higher commander immediately following receipt of an order) in that subordinate leaders are given time to complete their plans. (Refer to FM 6-0 for more information.) This ensures that unit plans are fully coordinated and in concert with the commander's intent. Commanders conduct confirmation briefs or backbriefs on a terrain model, a sand table, or a map. Planning for an airborne assault is a dynamic, fast-changing process. A change in one plan impacts other plans. Plans remain in draft until every commander in the chain has conducted a confirmation brief or backbrief. All commanders must inform their subordinates of changes.

3-33. Rehearsals are essential to the success of an airborne operation. They are conducted at every level, involve both air, and ground components. They are performed on terrain similar to the objective and under the same conditions. Rehearsals may be conducted on a sand table, terrain model, mock-up, or map, and if time permits a full-scale rehearsal. (See FM 6-0.) Rehearsals specific to airborne operations are listed in order of priority as follows:

- Ground tactical plan.
- Landing plan with emphasis to assemble on the drop zone.
- Air movement plan with emphasis on aircraft loading.

3-34. Leaders of the ABNAF must be able to make decisions to support the ABNTFC's intent. Plans and intelligence must be disseminated to the lowest level consistent with security requirements. (Refer to FM 2-0 for more information.) The staff follows security requirements in disseminating the intelligence required for subordinate units to develop their plans. Intelligence is provided on a need-to-know basis. As execution approaches, units are provided with more detailed intelligence. The commitment of an ABNAF is sudden and complete; there is no time for the commander to orient forces immediately after landing. Plans and intelligence must be thoroughly briefed before the operation begins.

WARNING ORDERS

3-35. To conduct an airborne assault, planning begins when the designated ABN IBCT receives a WARNORD from the ABNTF. The WARNORD specifies the ABNAFC and ABNTF task organization and allows the ABN IBCT staff to start initial planning and request supporting element liaison officers to report to the ABN IBCT headquarters early in the planning phase.

3-36. Once the ABNAFC receives the WARNORD, the planning process begins. This directive or WARNORD includes—

- Task organization.
- Mission command for the operation.
- Higher commander's concept of the operation (includes tentative scheme of maneuver/primary and alternate drop zones and landing zones).
- Missions for subordinate units.
- Time and duration of the operation (includes general timeline).

Airborne Assault Planning

- Intelligence and security requirements.
- Allocation and distribution of airlift assets.
- Unit deployment list and sequence.
- Departure airfields, remote marshalling bases, and intermediate staging bases.
- Initial estimate on requirements for airborne intelligence, surveillance, and reconnaissance; close air support; naval gun fire; and unmanned aircraft system support.
- Signal requirements and instructions.
- Linkup, withdrawal, and follow-on forces concepts.

3-37. Other WARNORDs and fragmentary orders (FRAGORDs) should follow as the ABNAF staff and commander work through the reverse planning sequence.

CONTROL MEASURES

3-38. ABNAFCs employ the full range of doctrinal control measures and graphics to delineate responsibilities, deconflict operations, safeguard friendly forces and civilians, and promote unity of effort. These measures include, but are not limited to boundaries that circumscribe operational area or area of operation, control measures to facilitate joint task force or ABNTF maneuver, fire support coordination measures (FSCMs); and airspace coordinating measures (ACMs).

AIRHEAD

3-39. An *airhead* is a designated area in a hostile or potentially hostile operational area that, when seized and held, ensures the continuous air landing of troops and materiel and provides the maneuver space necessary for projected operations (JP 3-18). Due to the nature of the airhead (a perimeter defense) and the required continuous airflow into the airhead, airspace coordinating measures and FSCMs must be established throughout the joint operational area, to include the ABNAF area of operation (drop zones, landing zones, assault objectives, and the airhead line).

3-40. During the initial stages of an airborne assault and before adequate ground communications can be established, coordination and control of fire support are accomplished from an airborne command and control platform. On landing, ABNTF and subordinate maneuver units establish contact with the airborne command and control platform through the tactical air control party (TACP) or fire support officer. Fire support, such as close air support, beyond that available from organic or direct support assets would be requested from the airborne platform. Prioritization and coordination of requests are accomplished by the ground force commander's representative in the airborne platform. Responsibilities include—

- Prevent fratricide of ground personnel.
- Ensure that requests do not interfere with incoming serials, other aircraft, or naval operations.
- Determine means of fire support coordination.
- Determine added safety or control measures required; transmit them to the appropriate ground elements.

3-41. Terminal guidance aids and control measures are used on the ground in the objective area to assist and guide incoming airlift aircraft to the designated drop zones and landing zones. Combat control teams comprised of USAF personnel are organized, trained, and equipped to provide aircraft terminal guidance. Army teams from the long-range surveillance company, a divisional or corps asset, are organized, trained, and equipped to deploy into the objective area and conduct reconnaissance and surveillance tasks before the deployment of the airborne force.

3-42. For airspace coordination, ABNAF staff establishes contact with the appropriate flight, provides essential information, and then hands the flight off to the appropriate TACP or forward air controller for mission execution. At that point, the mission is conducted the same way as conventional operations. If naval gun fire or air support is available, it is essential that a naval gunfire liaison officer be present in the airborne platform to perform a similar function.

3-43. The area air defense commander (AADC) is responsible for integrating the joint force air defense effort. All available surface-to-air assets should be incorporated into the overall air defense plan and comply with procedures and weapons control measures established by the AADC. The AADC exercises a degree of control of all systems through established guidelines, determines weapons control status, and joint force commander-approved procedural controls. (Refer to FM 3-01 and JP 3-01 for more information.)

3-44. Once adequate airspace ground control capabilities have been established in the airhead, fire support coordination responsibilities are passed from the airborne platform to the ground to be conducted as in other operations. (There is no doctrinal time for this transfer.) In some situations, this cannot occur; however, in most cases once a BCT main or tactical command post is on the ground, the transfer takes place.

PERMISSIVE AND RESTRICTIVE CONTROL MEASURES

3-45. Fire support coordination measures (FSCMs) both permissive and restrictive, are employed to ensure the safety of friendly personnel, to synchronize all fire support means, and to permit maximum flexibility with minimum restrictions on the employment of fire support. A common target and map grid system is established to permit transmission of target and friendly unit locations. This is critical if standard maps are not available. Provisions must be made to identify friendly force locations through the employment of smoke, panels, beacons, or other devices. (Refer to FM 3-90-1 for more information.)

DETERMINE GO/NO-GO CRITERIA

3-46. Abort criteria is a predetermined set of circumstances, based on risk assessment, which makes the success of an operation no longer probable; thus, the operation is terminated. These circumstances can relate to changes in safety, equipment or troops available, preparation or rehearsal time, weather, enemy, shaping operations prior to execution of the airborne assault, or a combination of the above. In the development of a course of action (during the preparation and deployment phase), airborne assault go/no-go criteria is developed. Criteria considerations include, but are not limited to:

- Minimum force:
    - Number of C130s or C17s.
    - Number lost of critical chalks, Infantry battalion equivalent, BCT assault command post, indirect fire systems, and mission command vehicle platforms.
    - Heavy-drop critical capability loss
- Intelligence/Pre-assault fires (yes/no):
    - Team recon no-fire areas established.
    - Light airfield repair package (LARP)— airfield damage is repairable with available equipment.
    - Enemy surface to air assets neutralized; self-propelled anti-aircraft gun on objective neutralized.
    - Indirect systems on objective destroyed.
    - Engineer forces on objective destroyed, or no larger than_____.
    - Air forces—air superiority achieved.
- Weather: Winds below 13 knots— Heavy equipment-17 knots—Personnel-13 knots.
- Fire support assets/airspace coordinating measures:
    - Close air support assets are on station with sufficient loiter time until P-hour +__ .
    - Electronic warfare assets on station.
    - Airspace coordinating measures active.
- Mission command sufficient assets until P-hour +____.

AIRFIELD OPENING

3-47. When developing an operation that may include an airfield opening, entry force tactical planners must have an understanding of the planning factors to consider for airfield opening in a hostile or permissive environment. Planners with specific airfield opening expertise to include the designation of a

Airborne Assault Planning

senior airfield authority are integrated as early as possible in the planning process. (Refer to FM 3-17.2 for more information.)

PLANNING AND ASSESSMENT

3-48. Planning for airfield opening begins at the strategic level where forces are assigned. Once employment forces are assigned by the joint task force, more detailed planning is conducted in coordination with the entry force and USAF contingency response group— first-responder airfield opening units. As combatant commanders identify airfields for use and direct their staffs to generate the appropriate plans, the missions and aircraft for which the airfield will be opened are determined and airfield capabilities are assessed.

3-49. Airfield assessment begins with airfield opening planning, which should begin as soon as the mission is assigned. Many tools are available to planners to begin the airfield assessment prior to actual arrival at the field. After arriving at the airfield, the airfield assessment team verifies the information gained during pre-mission planning with assault forces, collects additional data, and provides a recommendation to the airfield opening forces.

3-50. An airfield assessment should be accomplished rapidly to verify information and evaluate or obtain any items that were not pre-assessed. Assessments address areas such as runways, ramps, taxiways, force protection, communications, facilities, and provide a recommendation to appropriate decision makers on the suitability of future airfield operations (fixed- or rotary-wing).

OPERATION AND TRANSITION

3-51. Each contingency response group is scalable to meet specific tasking requirements. Primary capabilities/tasks include: airfield assessment, contingency response element command and control /port/quick-turn aircraft maintenance, force protection, intelligence, limited airfield security, airfield management and air traffic control, communications, fuels, medical, financial management, contracting, and supply.

3-52. When the responsibility for all or a part of an airfield changes from one organization to another, there is a requirement for a detailed and deliberate transfer. Likely transitions to occur during the life cycle of an airfield are—

- From airfield seizure to airfield opening.
- From airfield opening to follow-on or sustainment.
- From airfield sustainment to closure or turnover to the host nation.

SECTION IV – SHAPING OPERATIONS

3-53. Shaping operations establish conditions for the decisive operation through effects on the enemy, population, and terrain. Airborne operations may be designed as a sudden attack in force to achieve decisive results or as a shaping operation to create and preserve conditions for the success of a larger operation or campaign. When planning indicates the future requirement for an airborne assault, appropriate shaping operations or activities emphasize identifying and neutralizing an enemy's anti-access capabilities.

CREATE CONDITIONS

3-54. The ABNAF commander and his staff determine the exact conditions required according to the mission variables of mission, enemy, terrain and weather, troops and support available-time available, and civil considerations (METT-TC), to include the degree of acceptable risk with regard to each condition. Setting conditions is not limited to conducting SEAD and preparation fires. It requires the participation of numerous staffs, units, cells, and boards in different echelons and services.

3-55. Condition setting is an interactive process. The ABNAFC's situational understanding determines what part of the situation must change to ensure the success of the airborne assault. The ABNAFC tasks available reconnaissance forces and surveillance assets to detect the location of those enemy systems that unacceptably endanger the operation's success. This allows fire support systems to target and deliver effective fires against those enemy systems.

3-56. The most effective reconnaissance combines ground, aerial, and surveillance systems to provide constant coverage and multiple assessments of enemy activities throughout the objective area prior to the airborne assault. *Surveillance* is the systematic observation of aerospace, surface, or subsurface areas, places, persons, or things, by visual, aural, electronic, photographic, or other means (JP 3-0). The commander uses available reconnaissance forces and surveillance assets, to include available joint systems, to provide information that increases the accuracy of his situational understanding during planning and preparation.

3-57. The ABNAFC tasks the other warfighting functions to continue planning and preparing for the operation while employing service and joint fires to enable conditions. The ABNAFC requests assistance from higher echelons if there are not sufficient organic assets and information to accomplish the mission. The ABNAFC then assesses the progress of all the warfighting functions. This process repeats until the commander is satisfied with the set conditions or operational necessity forces him to either cancel or conduct the airborne assault.

3-58. Planning considerations should encompass special operations forces. Special operations forces may be inserted or already be operating in the objective area and become key components of the initial effort to shape and set conditions. Special operations forces regional expertise and environment preparation activities support well in advance of airborne assault planning and execution. Special operations forces may be introduced to the area well in advance of a possible assault to develop or prepare an area for airborne assault.

PRESERVE CONDITIONS

3-59. As the airborne assault extends in time and geography, extended lines of operations increase the assault unit's vulnerability to enemy capabilities designed to interrupt the expansion or reinforcement of the airhead or lodgment and follow-on operations. As the ABNAF expands its influence within an area of operation, the ABNAF becomes the primary means of setting conditions for operations that seize the initiative in other contested domains. The ABNAF leverages its presence to defeat enemy capabilities that limit freedom of action.

3-60. Continued high operational tempo and pressure preserves condition to hinder the enemy's ability to regroup, reconstitute capabilities, or reconfigure forces to support new plans. A primary means of maintaining continuous pressure is the continuous and rapid cycling of joint enablers and capabilities under operational level direction.

3-61. To rapidly transition from entry operations to follow-on operations, intermediate staging bases remain a critical part of baseline condition setting to: enable shorter range recycling of intra-theater lift capabilities, reorganization and reconfiguration of capabilities to meet evolving assault force demands and recalibration of battle and logistical rhythms.

Chapter 4
Ground Tactical Plan

The ground tactical plan is the base from which commanders develop all other plans. They must complete the ground tactical plan before finalizing the landing plan, the air movement plan, and the marshalling plan. It provides the commander's intent, his concept of the operation, fire support plan, and task organization of the units making the initial airborne assault. Ground combat following airborne operations is conducted along conventional lines but under unusual conditions. Once these conditions are appreciated, the tactics and methods of ground combat can be applied after the execution of airborne operations.

SECTION I – ELEMENTS

4-1. The ground tactical plan following an airborne assault contains essentially the same elements as other offensive operations. The elements, driven by the evaluation of the mission variables of METT-TC, are prepared to capitalize on speed and mobility to achieve surprise. Elements critical to the ground tactical plan include—

TASK ORGANIZATION

4-2. *Task organization* is a temporary grouping of forces designed to accomplish a particular mission (ADRP 5-0). Once ABNAFC determines the principal features of the ground assault plan (scheme of maneuver and fire support), he task organizes subordinate units to execute assigned missions and determines boundaries. To ensure unity of effort or to increase readiness for combat, part or all of the subordinate units of a command can be formed into one or more temporary tactical groupings (teams or task forces), each under a designated commander. Infantry units usually form the nucleus tactical groupings of the team; Infantry unit commanders lead the teams. These teams are tailored for the initial airborne assault by the attachment of required supporting units. They are attached as soon as possible in the marshalling area. Many of the units detach as soon as centralized control can be regained and the parent unit headquarters can be established on the ground. After the task organization of Soldiers for the airborne assault is announced, units organize into assault, follow-on, and rear echelons. (Refer to chapter 2 of this publication for more information.)

- Assault echelon. The assault echelon is composed of those forces required to seize the assault objectives and the initial airhead, plus their reserves and supporting Soldiers.
- Follow-on echelon. The airborne force does not need the follow-on echelon in the objective area during the initial assault, but does need it for subsequent operations. When needed, the follow-on echelon enters the objective area as soon as practical by air, surface movement, or a combination of the two.
- Rear echelon. The rear echelon includes the part of an airborne unit that is not considered essential for initial combat operations. It includes people left at its rear base to perform administrative and support functions that cannot be done efficiently in the combat area.

MISSION STATEMENT

4-3. The ABNAFC's mission statement is a short sentence or paragraph that describes the ABNAF's essential task (or tasks) and purpose— a clear statement of the action to be taken and the reason for doing so. The mission is analyzed in terms of the commander's intent two echelons up, mission statement (mission essential task and purpose) of the higher headquarters, specified tasks, and implied tasks. The

Chapter 4

mission of adjacent units must be analyzed to understand how they contribute to the decisive operation of their higher headquarters.

COMMANDER'S INTENT

4-4. The ABNAF *commander's intent* is a clear and concise expression of the purpose of the operation and the desired military end state that supports mission command, provides focus to the staff, and helps subordinate and supporting commanders act to achieve the commander's desired results without further orders, even when the operation does not unfold as planned (JP 3-0). ABNAF planners receive the ABNAF commander's intent as soon as possible after the mission is received. Even if the ground tactical plan is not complete, airborne assault planning often begins after the ABNAFC issues his intent.

CONCEPT OF OPERATIONS

4-5. The *concept of operations* is a statement that directs the manner in which subordinate units cooperate to accomplish the mission and establishes the sequence of actions the force will use to achieve the end state (ADRP 5-0). The concept of operations expands on the commander's intent by describing how the commander wants the force to accomplish the mission. It states the principal tasks required, the responsible subordinate units, and how the principal tasks complement one another. Commanders and staff use the operational framework to help conceptualize and describe their concept of operation.

4-6. The operational framework proves the commander with basic conceptual options for visualizing and describing operations in time, space, purpose, and resources. Commanders are not bound by any specific framework for conceptually organizing operations, and may use one of three conceptual frameworks listed below or in combination. These operational frameworks apply equally to both operational and tactical actions, and are listed as follows:
- The deep-close-security framework to describe the operation in time and space.
- The decisive-shaping-sustaining framework to articulate the operation in terms of purpose.
- The main and supporting efforts framework to designate the shifting prioritization of resources.

4-7. The deep-close-security operational framework has historically been associated with terrain orientation but can be applied to temporal and organizational orientations as well. Deep operations involve efforts to prevent uncommitted enemy forces from being committed in a coherent manner. Close operations are operations that are within a subordinate commander's area of operations. Security operations involve efforts to provide an early and accurate warning of enemy operations and to provide time and maneuver space within which to react to the enemy.

4-8. The decisive-shaping-sustaining framework lends itself to a broad conceptual orientation. The *decisive operation* is the operation that directly accomplishes the mission (ADRP 3-0). It determines the outcome of a major operation, battle, or engagement. A *shaping operation* is an operation that establishes conditions for the decisive operation through effects on the enemy, other actors, and the terrain (ADRP 3-0). A *sustaining operation* is an operation at any echelon that enables the decisive operation or shaping operation by generating and maintaining combat power (ADRP 3-0).

4-9. The main and supporting efforts operational framework—simpler than other organizing frameworks—focuses on prioritizing effort among subordinate units. Therefore, leaders can use the main and supporting efforts with either the deep-close-security framework or the decisive-shaping-sustaining framework. The *main effort* is a designated subordinate unit whose mission at a given point in time is most critical to overall mission success (ADRP 3-0). It usually is weighted with the preponderance of combat power. A *supporting effort* is a designated subordinate unit with a mission that supports the success of the main effort (ADRP 3-0). (Refer to ADRP 3-0 for more information.)

TASKS TO SUBORDINATES

4-10. Tasks to subordinate units direct individual units to perform specific tasks. They are a clearly defined and measurable activity accomplished by individuals and organizations and contribute to accomplishing the ABNAF mission or other requirements. The assignment of a task includes not only the task (what), but also the unit (who), place (where), time (when), and purpose (why). The purpose of each task should nest with completing another task, achieving an objective, or attaining an end state condition to the airborne assault. Example of activities include—

- Movement and maneuver. Maneuver units conduct an airborne assault and attack to destroy enemy forces on objectives. Units such as scouts, cavalry, long-range surveillance and special operations forces conduct reconnaissance and surveillance near the objective area, facilitate joint fires and close combat attack against identified enemy forces in the objective area, and conduct limited offensive tasks to interdict enemy forces.
- Intelligence. Ensures the information collection effort focuses on drop zones and landing zones and the objective area to identify enemy forces for targeting by fires and aviation assets to set conditions for airborne assault execution.
- Fires. While cannon artillery is part of the ABNAF, the primary support is close air support, naval gun fire and organic mortars initially on conduct of the airborne assault. Upon airland or follow-on force arrival, the field artillery battalion provides fire support on identified enemy positions on or near drop zones and landing zones to neutralize enemy forces and help set conditions for follow-on operations.
- Sustainment. Once the lodgment or airhead is secure for air-land or follow-on forces to arrive, forward logistics element from the brigade support battalion can begin casualty evacuation, resupply, equipment recovery, and refueling of vehicle and aviation assets in support of the ground tactical plan.
- Mission command. The ABNAFC may deploy in an airspace control aircraft to provide mission command oversight of the mission.

SECTION II – PLAN DEVELOPMENT

4-11. The ABNAFC begins to visualize the application of his ground tactical plan to his area of operation by defining the tactical problem and then begins a process of determining feasible solutions with his planning staff. The ground tactical plan incorporates considerations for those actions to be taken in the objective area, for example, during the assault and subsequent operations phases. This is the first plan to be finalized. It must be keyed on the accomplishment of the commander's concept of the operation.

4-12. The ground tactical plan is developed as other tactical plans using the procedure as delineated in FM 3-90.6, *Brigade Combat Team*. However, the initial goal of airborne operations is the establishment of an airhead and its subsequent defense. Essential elements of the ground tactical plan are developed in the following sequence:
- Assault objectives and airhead line (selected concurrently).
- Airhead and security area boundaries (developed sequentially).
- Assault force and security force task organization (developed sequentially).

MISSION VARIABLES OF METT-TC

4-13. When the ABNAF is alerted for deployment and assigned a mission, its assigned higher headquarters provides an analysis of the operational environment. That analysis includes the following operational variables: political, military, economic, social, information, infrastructure, physical environment, and time. The mission variables of METT-TC are used to filter the broader scope of operational variables into variables that directly affect a specific mission. The ABNAFC uses mission variables to gather relevant information for his mission analysis. This analysis enables him to combine operational variables and tactical-level information with knowledge about local conditions relevant to the mission. The following paragraphs address the mission variables of mission, enemy, terrain and weather, troops and support available-time available and civil considerations (METT-TC). (Refer to ADRP 5-0 for more information.)

MISSION

4-14. The mission of an airborne Infantry battalion or BCT is to close with the enemy by means of fire and movement to destroy or capture him, or to repel his assault by fire, close combat, and counterattack. These

Chapter 4

missions usually require the seizure and defense of objectives and surrounding terrain. Airborne assault forces rely strongly on the element of surprise.

ENEMY

4-15. Commanders analyze all available information to determine the enemy's situation. The following factors are considered:
- Enemy morale, leadership, and probable intentions.
- Enemy capabilities.
- Enemy tactics.
- Probable enemy reactions to an airborne assault.
  - The enemy that can react the fastest poses the immediate threat.
  - The enemy that can cause the most damage or prevent the airborne force from accomplishing its mission poses the most significant threat.
- Enemy reserves and paramilitary organizations (gendarmeries, police, border guards, and militia) and their ability to mobilize and react.
- Enemy capability to conduct guerrilla, partisan, or sabotage activities and the enemy's relationship to the local population.

TERRAIN AND WEATHER

4-16. The staff must consider these components; observation and fields of fire, avenues of approach, key terrain, obstacles, and cover and concealment (OAKOC) and then act on the following factors:
- The availability of drop zones and landing zones. Division or corps staff provides a landing area study to subordinate elements before the preparation of the airborne assault and follow-on landing plan. However, the availability and selection of drop zones should not influence the selection of assault objectives, the airhead line, or unit boundaries.
- Obstacles within the airhead line and out to the maximum effective range of direct- and indirect-fire weapons, with emphasis on those that can be prepared or reinforced with minimal engineer effort.
- Enemy avenues of approach, since the enemy tries to reach and destroy the airborne force before it can assemble and reorganize. This consideration weighs heavily in determining the location of assault objectives.
- Key terrain that can determine how the airborne force can best defend the area in-depth.
- Friendly and enemy observation and fields of fire (particularly for indirect fires and anti-armor weapons).
- Cover and concealment for movement and consolidation.
- The staff must consider the effects of climate and weather on—
  - Flight formations.
  - Trafficability.
  - Visibility.
  - Close air support.
  - Logistics.
  - Personnel and equipment.
  - Manned and unmanned aerial platforms.

TROOPS AND SUPPORT AVAILABLE

4-17. Commanders consider all forces available to accomplish the mission. These include all assigned, attached, and supporting forces.
- U.S. ground forces. Commanders evaluate the plans, missions, capabilities, and limitations of U.S. ground forces. They consider whether artillery can support the airborne forces and whether the forces perform a linkup or passage of lines.

- United States Air Force (USAF). Close air support often can compensate for the lack of armor and heavy artillery. The airborne commander must consider the Air Force's ability to support the force and must bring knowledgeable airlift and tactical air planners together early.
- United States Navy (USN). The airborne commander examines the availability and feasibility of naval gun fire support and naval or U.S. Marine Corps (USMC) air support. Early arrangements for liaison and coordination must be made to support the operation.

TIME AVAILABLE

4-18. Time is critical in all operations. There are several time considerations that are unique to an airborne operation. Significant time may be required to mass the lift force. The time between the initial assault and the deployment of the follow-on echelon must be considered. The amount of time before linkup or withdrawal drives sustainment planning.

CIVIL CONSIDERATIONS

4-19. Understanding the operational environment requires understanding the civil aspects of the joint operational area. Social and economic variables often receive close analysis as part of civil considerations at brigade and higher levels. Depending on mission, the ABNTFC considers national and regional characteristics such as—
- Religion and customs.
- Politics and tribal affiliations.
- Support or lack of it for central and local governments or occupying powers.
- Loyalty to political or military leaders.
- Available labor.
- Support or lack of it for U.S. forces.

4-20. *Civil considerations* is the influence of manmade infrastructure, civilian institutions, and activities of the civilian leaders, populations, and organizations within an area of operations on the conduct of military operations (ADRP 5-0). The ability to analyze civil considerations to determine their impact on operations at brigade and below enhances several aspects of the airborne operation to include insertion into the objective area, seizure of assault objectives and establishment of the airhead, and follow-on operations. Civil considerations comprise six characteristics, expressed in the memory aid ASCOPE—areas, structures, capabilities, organizations, people, and events. (Refer to ATP 2-01.3 for more information.)

ASSAULT OBJECTIVE AND AIRHEAD LINE

4-21. Based on his analysis of METT-TC, the commander selects specific assault objectives. (See figure 4-1, page 4-6.) Although the airhead line is developed and the assault objectives determined concurrently, the assault objectives dictate the size and shape of the airhead.

Chapter 4

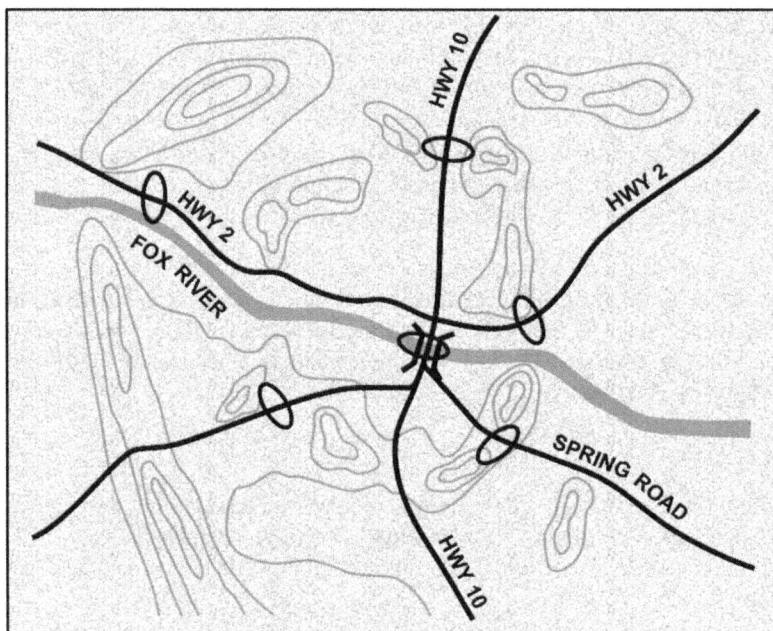

Figure 4-1. Assault objectives

4-22. This selection does not necessarily include those objectives that the force must seize to secure the airhead line. An appropriate assault objective is one, which the force must control early in the assault to accomplish the mission, or it must control to enhance the security of the airborne unit during the establishment of the airhead.

4-23. Objectives should allow for the accomplishment of mission-essential tasks while meeting the commander's intent. They can include key terrain within the airhead or terrain required for linkup. For example, the commander has directed the airborne force to secure a bridge for later use by linkup forces. The force must secure this bridge before the enemy can destroy or damage it; therefore, the commander designates the bridge as an assault objective.

4-24. The airborne unit is vulnerable from the time of the airborne assault until follow-on forces can be delivered to the airhead. A mobile enemy unit attacking the airhead during these early moments can completely disrupt the operation. Therefore, the commander selects assault objectives terrain that dominates places where high-speed enemy avenues of approach enter the airhead.

4-25. Enemy positions that both threaten the mission and are located within the airhead can be selected as assault objectives. However, commanders would not classify mobile forces as assault objectives.

4-26. Assault objectives must be seized immediately to establish the airhead and to provide security for follow-on forces coming into the airhead.

4-27. Other considerations influence the development and final selection of assault objectives. Subordinate commanders decide the size, type, or disposition of the force to gain/maintain control.

- Division selects brigade assault objectives.
- Brigade selects battalion assault objectives.
- Battalion selects company assault objectives.
  - Senior commanders choose as few assault objectives as possible since subordinate commanders must select additional objectives to establish a cohesive defense of their assigned areas of the airhead.
  - Assault objectives are ranked in order. A unit SOP may predesignate a numbering system for subordinate objectives. For example, all first brigade objectives begin with a "Q," or for OPSEC purposes, they may be randomly numbered or lettered. Priorities are chosen based on the most likely threat or on the needs of the friendly force.

- Assault objectives are secured before the defense is setup in the airhead line. The airhead is then cleared of organized enemy resistance and forces are positioned to secure the airhead line.

4-28. When commanders select assault objectives, they should consider the extent of the airhead. The airhead includes the entire area under control of the airborne force. It acts as a base for further operations and as a respite that allows the airborne force to build combat power. Once the force secures the airhead, they must clear enemy forces within it; then, they must defend it.

4-29. The *airhead line* is a line denoting the limits of the objective area for an airborne assault (JP 3-18). It delineates the specific area to be seized and designates the airhead. Several principle factors determine the location, extent, and form of the airhead or airhead line, they are as follows:
- The actual trace of the airhead line reflects the control of key or critical terrain essential to the mission. (See figure 4-2.)

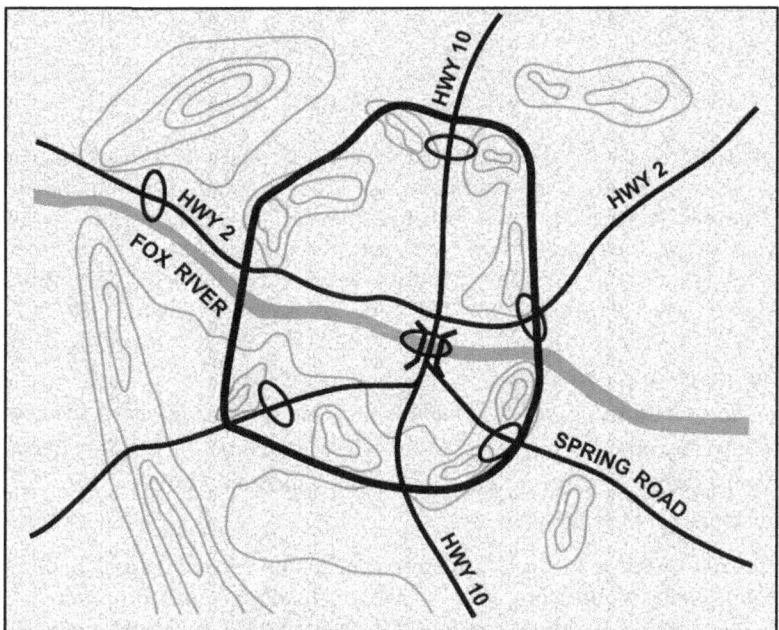

Figure 4-2. Airhead line

- The airhead line anchors on obstacles, and the airhead itself takes advantage of natural and man-made obstacles.
- The airhead contains enough drop zones and landing zones to ensure interior rather than exterior lines of communication and to permit mass rather than piecemeal assault.
- The airhead must allow enough space for dispersion to reduce vulnerability to chemical, biological, radiological, nuclear weapons.
- The airhead must be large enough to provide for defense in depth, yet small enough for the unit to defend. Although this is largely METT-TC dependent, a battalion can defend an airhead three to five kilometers in diameter. An IBCT can occupy an airhead five to eight kilometers in diameter.

SECURITY AND RECONNAISSANCE

4-30. Security in all directions is an overriding consideration early in the airborne operation, since an airhead or lodgment is essentially a perimeter defense. Another overriding consideration, the airborne unit's method of arriving into the objective area requires the conduct of an immediate and thorough reconnaissance and the rapid transmission of this information to higher headquarters. Security operations and reconnaissance missions within the security area enable these efforts.

Chapter 4

4-31. *Security operations* are those operations undertaken by a commander to provide early and accurate warning of enemy operations, to provide the force being protected with time and maneuver space within which to react to the enemy, and to develop the situation to allow the commander to effectively use the protected force (ADRP 3-90). *Reconnaissance* is a mission undertaken to obtain, by visual observation or other detection methods, information about the activities and resources of an enemy or adversary, or to secure data concerning the meteorological, hydrographic, or geographic characteristics of a particular area (JP 2-0).

4-32. After the force makes the initial assault landing and accomplishes its first missions, the commander organizes the airhead line as the defense perimeter. The terrain and situation dictates how units occupy and organize the airhead line. Forces assigned reconnaissance and security tasks, usually include reinforcing the security area. The mission, enemy capabilities, and defensive characteristics of the terrain determine the degree to which the airhead line is actually occupied and security area forces are organized for the airhead. Task organization of security and reconnaissance forces is METT-TC-dependent and may include scouts, Infantry, cavalry, antitank weapons, engineers, Army aviation, electronic warfare, and follow-on Stryker and Armored forces.

Security Forces

4-33. Security forces land early in the assault echelon. In the early stages of an airborne operation, the security force acts as a screening force. In later stages (when assault missions have been accomplished, when the airhead is relatively secure, and when more forces are available), it acts as a guard force. The security area is established four to six kilometers from the airhead line to afford security to the airborne force during its landing and reorganization. Security forces come under BCT control except during short missions such as raids, when they come under battalion or squadron control. The mission of the security force is to—

- Give the airhead early warning.
- Develop information collection, to include the location, direction, and speed of an enemy attack.
- Deny the enemy observation of and ability to direct indirect fire on the airhead.
- Deceive the enemy as to the actual location of the airhead.
- Delay and disrupt the enemy.

4-34. The need for and positioning of additional security forces is determined by the next subordinate commander. When possible, mobile forces are selected to facilitate rapid initial movement to positions and to facilitate withdrawal and adjustment. An aggressive reconnaissance and surveillance effort at lower echelons augments the security force. The following considerations apply to the selection of positions for the security force:

- Locate them within radio communications and fire support range. However, this range can be extended, if necessary, with retransmission stations, split section indirect-fire operations, and attachment of vehicles, mortars, or other assets to the security force.
- Locate them as roadblocks, obstacles, ambushes, patrols, or sensors (depending on the enemy) on dominant terrain. This allows long-range observation and fields of fire out to the maximum range of supporting fires.
- Locate them to observe, control, and dominate enemy high-speed avenues of approach.
- Locate them to deny enemy long-range observation and observed indirect fire into the airhead.
- Locate them far enough out to provide early warning.
- Locate them to provide routes of withdrawal to the airhead. Observation posts generally rely on their ability to hide as their main protection; they can allow the enemy to pass their position and not withdraw.

Reconnaissance Forces

4-35. Designated forces under control of the ABNAFC perform reconnaissance missions within the security area; emphasis is placed on likely enemy avenues of approach. The mission of these forces is to gain and maintain contact with enemy units reacting to the airborne assault. This force is mobile and not

used to defend a particular part of the airhead. It may be supported with fire from USAF assets, naval gun fire, or Army missile systems. The following considerations govern the employment of this force:
- These forces orient on enemy high-speed avenues of approach to develop intelligence to include the location, direction, and speed of the enemy's advance.
- Commanders of these forces consider known enemy locations, the number of high-speed approaches, and communications-relay abilities while orienting on enemy units.
- Usually employed beyond the airhead at a distance based on the tactical situation, commanders can extend their range if communications permit.
- Aviation assets can extend to 50 kilometers or more, although the commander must consider loiter time so the forces can provide continuous coverage. (Forward arming and refueling points [FARPs] can increase this distance.)
- Long-range surveillance teams may perform surveillance of enemy garrisons and major routes into the airhead.
- Reconnaissance forces must be mobile and task-organized for the mission.

BOUNDARIES

4-36. Commanders visualize the employment of subordinate units to organize them for combat commensurate with the missions. Commanders use boundaries to assign areas of responsibility to major subordinate combat elements, which then clear the area of enemy forces. (See figure 4-3, page 4-10.) In selecting and designating assault boundaries for airborne operations, several points are considered as follows:
- Each unit should be able to clear its assigned area; therefore, commanders must consider boundaries concurrently with task organization. To assign boundaries, commanders subdivide the area into areas with equal tasks (not necessarily into equal areas). This requires a careful analysis of the enemy, the tasks to be accomplished, and the terrain within the objective area.
- Commanders should avoid splitting (between two units) the responsibility for the defense of an avenue of approach or key terrain.
- Commanders should ensure there is adequate maneuver space in the area, to include key terrain features that control it.
- Commanders should avoid designating boundaries in such a way that a major terrain obstacle divides a unit area; this presents problems for maneuvering forces.
- The boundaries should provide adequate room to permit the commander to maneuver forces on both sides of their assault objectives.
- Commanders must choose boundaries that are recognizable both on the map and on the ground. Roads should not be used as a boundary because they represent a high-speed avenue of approach and need to be covered with a clear understanding of responsibility. Instead, commanders can use rivers, streams, railroad tracks, the edge of a town, woods, the edge of a swamp, and so on.
- An ABN IBCT area of operation should include at least one drop zone and one landing zone to allow for follow-on forces to land during the assault. This facilitates resupply and evacuation of enemy prisoners of war and casualties. Having a drop zone and landing zone allows for ease of sustainment operations. Regardless of boundaries, units should drop on the drop zone closest to their assault objective.
- Commanders should establish boundaries that serve during the airborne assault and during later operations. These should be readily recognizable during limited visibility.
- Commanders should choose boundaries that do not require a unit to defend in more than one direction at once. They should not expect a unit to secure objectives within the airhead at the same time they establish its defense.
- Boundaries should extend beyond the trace of the security force as far as needed to coordinate fires. This allows subordinate units to operate forward of the airhead with minimal coordination.
- Commanders should plan coordinating points at the intersection of the airhead line and security force ground trace boundaries.

Chapter 4

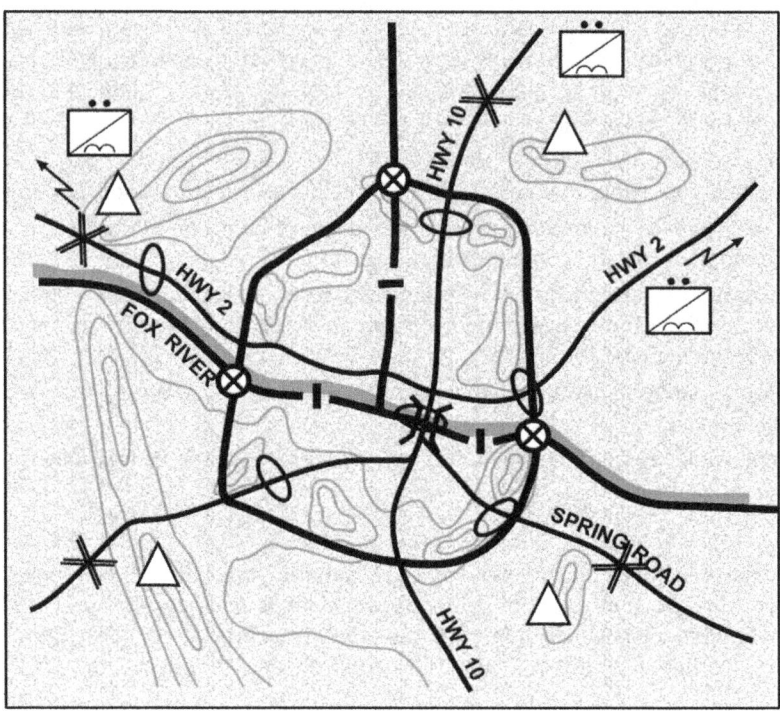

Figure 4-3. Boundaries

METHOD OF ATTACK

4-37. There are three basic methods of attacking an objective. They are:
- Jumping or landing on top of the objective. This method works best for attacking a small objective that is specially fortified against ground attack. However, an airborne landing into an area strongly defended against air attack requires surprise to succeed.
- Jumping or landing near the objective. This method works best for the capture of a lightly defended objective that must be seized intact such as a bridge. If the enemy has strong defenses against air attack, only surprise can enable the unit to achieve success with few casualties.
- Jumping or landing at a distance from the objective. This method is the least often used of the methods available. Airborne forces use this method for large complex objectives that must be seized by deliberate attack. The drop zone is selected to emphasize security and preservation of the force. The plan is based on METT-TC considerations and should surprise the enemy.

METHOD OF LANDING

4-38. There are two basic landing methods. They are as follows:
- Multiple drop zones. The use of multiple drop zones creates a number of small airheads in the objective area. This method supports the principle of mass by placing the maximum number of paratroopers in the objective area in the minimum amount of time. Additionally, the commander can capitalize on the principle of surprise because the main effort is not easily determined by the enemy. This method normally is used by division-size elements and larger.
- Single drop zone. IBCT and smaller-size airborne forces often establish an airhead by conducting the airborne assault onto a single drop zone. This method allows the assaulting unit to assemble quickly and mass combat power against the enemy.

TIME-SPACE FACTORS

4-39. Commanders schedule the delivery sequence and the time between serials to provide the least time and distance separation between each aircraft and serial. The airborne force assembles maximum combat power on the drop zone as quickly as possible, using either of the following options:
- Land all elements in the same area. Aircraft approach the drop zone in a deep, narrow formation and all Soldiers jump into a small area.
- Land all elements at the same time. Aircraft in a wide formation approach various drop zones situated close to each other and all Soldiers jump at the same time or as near to it as possible.

LANDING PRIORITIES

4-40. Airborne units are cross loaded to land close to their assault objectives and are organized to try to maintain tactical unity.
- Battalions or battalion task forces land intact on a single drop zone. An IBCT lands in mutually supporting drop zones. Two or more battalions land successively on the same drop zone or each can land on a separate drop zone within a general IBCT drop zone area.
- The airborne force sends as many assault unit personnel and equipment as possible into the area in parachute serials. Commanders must consider the mobility of equipment after the landing. For example, the carriers or prime movers that are deliverable by parachute, but difficult to move or carry on the ground; can accompany the weapons in the assault element. Paratroopers accompany their units' principal items of equipment.

FIRE SUPPORT PLAN

4-41. The following fire support planning and coordination actions are the responsibility of the IBCT/battalion/squadron during ground movement:
- Support the scheme of maneuver. The goal is to place the maximum amount of indirect firepower on the ground as quickly as possible.
- Control indirect-fire systems. Initially, control is decentralized; a forward observer calls for fire directly to a fire support asset.
- Plan fires to block enemy avenues of approach. (Consider family of scatterable mines [FASCAM] delivered by air.)
- Plan fires to eliminate enemy resistance (groups and series in the objective area).
- Plan fires to defend key terrain needed to link up with friendly forces.
- Plan fires to support security or reconnaissance forces in the objective area.
- Plan fires on top of, to the flanks, and beyond assault objectives.
- Plan close air support.
- Plan final protective fires.
- Recommend priority of fires.
- Select initial field artillery and mortar positions that can be occupied quickly from drop zones and landing zones.
- Select subsequent field artillery and mortar positions to provide combat outposts and security forces.
- Coordinate landing plan. Planning and coordination of fire support during the air movement and pre-assault fires are the joint task force's responsibility; he plans SEAD fires along the flight route and in the objective area. Once on the ground, friendly positions are marked. The assault force must ensure that pre-assault air strikes are planned against other enemy positions in the objective area.
- Pre-assault fires are planned as follows:
  - On and around the landing zone or drop zone (alternate and false).
  - On enemy air defense artillery.
  - On enemy command, control, and communication.

- On enemy indirect-fire systems.
- Sequence and location of delivery for field artillery and mortars.

SECTION III – AIR-GROUND OPERATIONS

4-42. Air action by fixed- and rotary-wing aircraft against hostile targets that are in close proximity to friendly forces require detailed integration of each air mission with the fire and movement of ground forces. This section discusses fundamental considerations for effective air-ground operations, close combat attack, and close air support. It addresses unmanned aircraft system operations to provide surveillance capabilities and to enhance the ABNAFC's situational awareness as he plans, coordinates, and executes the airborne assault.

FUNDAMENTAL CONSIDERATIONS

4-43. To ensure effective air-ground operations, airborne assault commanders and staffs must consider the integration of air and ground maneuver forces. The following fundamentals provide a framework for enhancing the effectiveness of both air and ground maneuver:
- Understanding capabilities and limitations of each force.
- Using standard operating procedures.
- Forming habitual relationships.
- Using regular training events.
- Rehearsals.
- Maximizing and concentrating effects of available assets.
- Synchronization.

4-44. Integration involves merging the air and ground fights into one to apply proper aviation capabilities according to the supported ABNAFC's intent. Integration ideally begins early in the planning process with the involvement of the air defense airspace management/brigade aviation element (ADAM/BAE). The ADAM/BAE advises the ABNAFC on aviation capabilities and the best way to use aviation to support mission objectives. Ensuring the AVN LNO or BAE passes along the task and purpose for aviation support and continually provides updates as needed is of equal importance. Simply stated, ensuring the aviation brigade and subordinate unit staffs fully understand the ABNAF scheme of maneuver and commander's intent is critical to successful air-ground operations.

CLOSE COMBAT ATTACK

4-45. A *close combat attack* is a coordinated attack by Army attack reconnaissance aircraft (manned and unmanned) against enemy forces that are in close proximity to friendly forces. The close combat attack is not synonymous with close air support flown by joint aircraft. Terminal control from ground units or controllers is not due to the capabilities of the aircraft and the enhanced situational understanding of the aircrew (FM 3-04.126). In most instances, the attack aviation already may occupy holding areas, battle or support by fire positions or are in overwatch of the ground unit as it begins its assault. The ABNAF employs close combat attack procedures to ensure that these aviation fires destroy the enemy with minimal risk to friendly forces.

4-46. After executing the airborne assault, employing attack reconnaissance aviation with ground maneuver forces requires coordinated force-oriented control measures and the close combat attack (CCA) 5-Line attack brief allowing aviation forces to support ground maneuver with direct fires while minimizing fratricide risks. The aviation liaison officer should identify early in the planning process the minimum ABNAF graphics required for operations (boundaries, phase lines, attack by fire positions, objectives, and so on). Brigade aviation elements and liaison officer personnel should ensure that supported units are familiar with close combat attack request procedures and marking methods

CLOSE COMBAT ATTACK REQUEST

4-47. A close combat attack is coordinated and directed by a team, platoon, or company level ground unit using the standardized Format 22. CCA 5-Line attack brief. (Refer to chapter 10, section III of this publication for more information.) The most important factor of successful close combat attack is positive and direct communication between aviation and ground forces. As stated earlier, close combat attack does not require a joint terminal attack controller (JTAC) unlike close air support missions. However, utilizing a joint fires observer can minimize the risk of fratricide and expedite the clearance of fires procedures. (Refer to FM 3-04.126 and ATP 3-09.32 for more information.)

4-48. Any element in contact uses the CCA 5-Line attack brief to initiate the close combat attack. The CCA 5-Line attack brief allows the ground maneuver forces to communicate and reconfirm to the aircraft the exact location of friendly and enemy forces. The ground commander owning the terrain clears fires during the close combat attack by giving aircrews the situational awareness of the location of friendly elements. The ground commander deconflicts the airspace between indirect fires, close air support, and the close combat attack aircraft.

4-49. After receiving the request for close combat attack, the aircrew informs the ground unit leader of the battle position, attack or, support by fire position (or series of positions) the team is occupying, and the location from which the attack aircraft engages the enemy with direct fire. The size of this position varies depending on the number of aircraft using the position, the size of the engagement area, and the type of terrain. The position must be close enough to the requesting unit to facilitate efficient target handover. Aircraft leaders normally offset the position from the flank of the friendly ground position. This helps to ensure that rotor wash, ammunition-casing expenditure, and the general signature of the aircraft do not interfere with operations on the ground. The offset position allows the aircraft to engage the enemy on his flanks rather than its front. It reduces the risk of fratricide along the helicopter gun-target line.

TARGET HANDOVER

4-50. The rapid and accurate marking of a target is essential to a positive target handover. Aircraft conducting close combat attacks develop an attack plan that is METT-TC dependent and meets the ground commander's task and purpose. The aircrew generally has an extremely limited amount of time to acquire both the friendly and enemy locations. It is essential that the ground unit has the marking ready and turned on when requested by the aircrew. Attack reconnaissance aircrews use both thermal sight and night vision goggles (NVGs) to fly with and acquire targets. After initially engaging the target, the aircrew generally approaches from a different angle for survivability reasons if another attack is required. The observer makes adjustments using the eight cardinal directions and distance (meters) in relation to the last round's impact and the actual target. At the conclusion of the close combat attack, the aircrew provides its best estimate of battle damage assessment to the unit in contact.

BATTLE DAMAGE ASSESSMENT AND REATTACK

4-51. After the attack aircraft complete the requested close combat attack mission, the aircrew provides a battle damage assessment to the ground commander. Based on his intent, the ground maneuver commander determines if another attack is required to achieve his desired end state. The close combat attack operation can continue until the aircraft have expended all available munitions or fuel. However, if the air mission commander receives a request for another attack, he must carefully evaluate his ability to extend the operation. If not able, he calls for relief on station by another attack team if available. It is unlikely that the original team has enough time to refuel, rearm, and return to station.

CLEARANCE OF FIRES

4-52. During an airborne assault with numerous aircraft in the vicinity of the drop zone, it is critical to deconflict airspace between aircraft and established indirect fires once air-land or follow-on forces arrive, to include the following:
- Ensure aircrews have the current and planned indirect fire positions (to include mortars) supporting the ground tactical plan.
- Plan for informal airspace coordination areas and check firing procedures and communications to ensure artillery and mortars firing from within the drop zone do not endanger subsequent serials landing or departing, close combat attack, or close air support.

Chapter 4

- Ensure that at least one aviation team members monitors the fire support network for situational awareness.
- Advise the aviation element if the location of indirect fire units changes from that planned.
- Ensure all participating units are briefed daily on current airspace control order or air tasking order changes and updates that may affect air mission planning and execution.
- Ensure all units update firing unit locations, firing point origins, and final protective fire lines as they change for inclusion in current airspace control order.

4-53. The ABNAFC or ground commander can establish an airspace coordination area. For example, he can designate that all indirect fires be south of and all aviation stay north of a specified gridline for a specific period. This is one method for deconflicting airspace while allowing both indirect fires and attack aviation to attack the same target. The ground commander then can deactivate the informal airspace coordination area when the situation permits.

CLOSE AIR SUPPORT

4-54. *Close air support* is air action by fixed- and rotary-wing aircraft against hostile targets that are in close proximity to friendly forces and that require detailed integration of each air mission with the fire and movement of those forces (JP 3-0). Like close combat attack, close air support can be conducted at any place and time friendly forces are in close proximity to enemy forces based on availability. All leaders in the ABNAF should understand how to employ close air support to destroy, disrupt, suppress, fix, harass, neutralize, or delay enemy forces. (Refer to JP 3-09.3 for more information.)

4-55. Only joint terminal attack controllers (JTACs) or forward air controllers (airborne) (FAC[A]s) personnel have the authorization to perform terminal control of close air support aircraft during operations (combat and peacetime) within proximity of their supported ground combat units. Nomination of close air support targets is the responsibility of the commander, air liaison officer, and S-3 at each level. The ABNAF may receive close air support from USAF, USN, USMC, or multinational units. (Refer to ATP 3-09.32 for more information.)

CAPABILITIES AND EMPLOYMENT

4-56. Before and during an airborne assault, USAF aircraft are available to provide close air support. Requests for these aircraft are processed through the tactical air control party (TACP) colocated with the IBCT main command post. The TACP is organized as an air execution cell capable of requesting and executing Type 2 or 3 terminal attack control of close air support missions. The staffing of the cell depends on the situation but at a minimum, includes an air liaison officer and a JTAC. (Refer to ATP 3-09.32 for more information.) To use close air support aircraft, the leader on the ground should be familiar with the characteristics of the aircraft predominantly used in the close air support role.

BRIEFING FORMAT

4-57. Two types of close air support requests are as follows:
- Preplanned requests that may be filled with either scheduled or on-call air missions. Those close air support requirements foreseen early enough to be included in the first air tasking order distribution are submitted as preplanned air support requests for close air support. Only those air support requests submitted in sufficient time to be included in the joint air tasking cycle planning phases and supported on the air tasking order are considered preplanned requests.
- Immediate requests that are mostly filled by diverting preplanned missions or with on-call missions. Immediate requests arise from situations that develop outside the air tasking order planning cycle.

4-58. The air liaison officer and JTAC personnel in the TACP are the primary means for requesting and controlling close air support. However, reconnaissance units conducting shaping operations, such as reconnaissance and surveillance missions that have joint fires observer certified personnel, may observe and request close air support through the JTAC. (Refer to ATP 3-09.32 for more information.)

Ground Tactical Plan

UNMANNED AIRCRAFT SYSTEM

4-59. Before the airborne assault, the ABNTFC may rely on unmanned aircraft system (UAS) operations to provide surveillance capabilities and to enhance the ABNAFC's situational awareness as he plans, coordinates, and executes the airborne assault. UAS employment prior to the assault is weighted against the element of surprise. Once the airborne assault is executed, the commander can employ UAS from his organic elements or he can request to have direct access to real-time feeds from additional UAS support from his higher headquarters. UAS employment is particularly effective when employed together with ground and attack reconnaissance elements as a team during shaping operations in which the commander is trying to create the conditions for successful airborne assault execution. (Refer to FM 3-04.155 for more information.)

CAPABILITIES

4-60. UAS bring numerous capabilities to the ABNAF. Employment of these systems before executing the airborne assault and employment with ground and attack reconnaissance units provides reconnaissance, surveillance, and target acquisition capabilities. UAS also can support military deception by flying in an area to make the enemy think it is a friendly objective. The RQ-7B Shadow can participate in attack operations by either employing indirect fires and by laser designation of targets for joint aircraft and remote engagements by armed manned and unmanned aircraft. The MQ-1C Gray Eagle can do the same and may be armed with Hellfire missiles to engage autonomously or fire its missiles for a remote designator.

Reconnaissance Operations

4-61. When UAS complement the ground reconnaissance units during reconnaissance operations, they operate forward of the element (METT-TC dependent). They can conduct detailed surveillance of areas that are particularly dangerous to ground reconnaissance units, such as drop zones, landing zones and objective areas. They can be employed effectively in support of operations in urban terrain.

4-62. They can support route reconnaissance forward of reconnaissance and security units or be employed in conjunction with reconnaissance and security units when it is necessary to reconnoiter multiple routes simultaneously. The reconnaissance unit leader can employ UAS to support an area or zone reconnaissance mission. Upon contact, UAS provide early warning for the element and then maintain contact until the element conducts a reconnaissance handover from the UAS to another element.

Security Operations

4-63. In security operations, UAS complement reconnaissance units by assisting in identification of enemy reconnaissance and main body elements and by providing early warning forward of reconnaissance units. Besides acquiring enemy forces, UAS can play a critical role in providing security through the depth of the screen by observing dead space between ground observation posts. They can support reconnaissance units during area security missions by screening or conducting reconnaissance.

Reconnaissance/Target Handover

4-64. When a UAS makes contact, particularly during reconnaissance operations, the operator hands over the UAS contact to ground or attack reconnaissance units as quickly as possible. Rapid handover allows the UAS to avoid enemy air defense weapons and helps maintain the tempo of the operation. During the handover, the UAS assists in providing direction to the ground or attack reconnaissance unit charged with establishing contact with or engaging the enemy. It maintains contact with the enemy until the units are in position and have established sensor or visual contact.

4-65. The first action in the handover process is a report (such as, spot report or situation report) from the UAS operator to the ground or attack reconnaissance unit. Next, the UAS reconnoiters the area for secure positions for the unit (such as hide, overwatch, observation posts or battle positions) and likely mounted and dismounted routes into the area. The ground or attack reconnaissance unit moves to initial hide positions along the route selected by the leader based on UAS-collected information. The ground or attack reconnaissance unit then moves to establish sensor or visual contact with the enemy. Once this contact is established, the ground or attack reconnaissance unit sends a report to the UAS operator. When the UAS operator confirms that the ground or attack unit can observe enemy elements and has a clear picture of the situation, handover is complete. The UAS then can be dedicated to another mission or, in the case of target handover to attack reconnaissance units, may be used for battle damage assessments and reattack if necessary.

SECTION IV – EXECUTION

4-66. The initial airborne assault emphasizes the coordinated action of small units to seize initial battalion objectives before the surprise advantage has worn off. As assault objectives are seized, the airborne force directs its efforts toward consolidating the airhead.

CONDUCT OF THE AIRBORNE ASSAULT

4-67. Tactical surprise and detailed planning should enable units to seize their assault objectives and to establish the airhead before the enemy has time to react in force. Missions of units change when necessary by the enemy defense of initial objectives. The enemy can be expected to launch quick uncoordinated attacks along major avenues of approach using local forces. The degree of coordination and strength of these attacks increase progressively, therefore, the airborne force must develop correspondingly greater strength in its defensive positions. Preparing early defense against armored attack is a major consideration.

4-68. Units assigned to perform reconnaissance and security missions should be cross loaded in the load plan so that during the airborne assault, they are to be some of the first elements on the ground to move and establish roadblocks, locate enemy forces; disrupt enemy communication facilities; and provide the commander with early warning, security, and information. Since ground reconnaissance by unit commanders is seldom possible before the airborne operation, it must begin immediately after units hit the ground, and assemble. The information flow must be continuous. Information requirements do not vary from those employed by other ground units. However, the unit's method of arrival in the area of operations makes immediate and thorough reconnaissance and transmission of information to higher headquarters necessary.

4-69. If the initial objectives are heavily defended, the bulk of the force is assigned the task of seizing these objectives. When initial objectives are lightly defended, the bulk of the force can be employed in clearing assigned areas and preparing defensive positions in depth. Extensive patrolling is initiated early between adjacent defensive positions within the airhead line, and between the airhead and the limits of the security area. Army aircraft are well-suited for support of this patrolling effort. Contact with friendly guerrilla forces, long-range surveillance teams and special operations forces in the area is established as soon as possible.

4-70. Brief personnel on unit plans, adjacent and higher units' plans, and alternate plans. This helps units or personnel landing in unplanned areas to direct their efforts to accomplishing the mission. Misdelivered units or personnel establish contact with their respective headquarters as soon as practical.

4-71. Sufficient communication personnel and equipment must be moved into the airhead in advance of the command post they are to serve to ensure the timely installation of vital communication. (Refer to FM 6-02 for more information.) As soon as communication and the tactical situation permit, commanders regain centralized mission command. Therefore, immediate establishment of the following is necessary for effective mission command:
- Command and fire control channels within the airborne forces.
- Communication with supporting air and naval forces.
- Communication with airlift forces concerned with buildup, air supply, and air evacuation.
- Communication with bases in friendly territory.
- Communication between widely separated airborne or ground forces, such as linkup forces, with a common or coordinated mission.

4-72. The commander influences the action by—
- Shifting or allocating fire support means.
- Moving forces.

- Modifying missions.
- Changing objectives and boundaries.
- Employing reserves.
- Moving to a place from which he can best exercise personal influence, especially during the initial assault.

4-73. When initial objectives have been secured, subordinate units seize additional objectives that facilitate the establishment of a coordinated IBCT defense or the conduct of future operations. Defensive positions are organized, communications are supplemented, and reserves are reconstituted. These and other measures are taken to prepare the force to repel enemy counterattacks, to minimize the effects of attack by chemical, biological, radiological, and nuclear means, or to resume the offensive.

4-74. Reserves prepare and occupy blocking positions, pending commitment. Typical missions for reserves committed during the initial assault include taking over the missions of misdelivered units, dealing with unexpected opposition in seizing assault objectives, and securing the initial airhead.

DEVELOPMENT OF THE AIRHEAD

4-75. After the airborne forces make the initial assault landings and accomplish the initial ground missions, commanders must organize the airhead line. Considerations include:
- Size. The airhead line extends far enough beyond the landing area to ensure uninterrupted landings of personnel, equipment, and supplies. It secures the requisite terrain features and maneuver space for such future offensive or defensive tasks as called for in the mission.
- Occupation and organization. Units occupy and organize the airhead line to the extent demanded by the situation. Commanders adjust the disposition of units and installations to fit the terrain and the situation. Units take reconnaissance and security measures; this usually includes the reinforcement of the security area. The degree to which the airhead line is actually occupied and organized for defense is largely determined by the mission, enemy capabilities, and the defensive characteristics of the terrain.
- Buildup. This proceeds concurrently with seizing and organizing the airhead line and, if required, repairing or constructing an airfield to receive follow-on airland echelons. As more combat personnel arrive and commanders organize them by unit, positions are reinforced on and within the airhead line, reserves are constituted, and preparations are made for such offensive tasks as the mission requires.

BUILDUP OF COMBAT POWER

4-76. The buildup of combat power is the introduction of the follow-on echelon into the airhead. This increase of friendly combat power yields a corresponding ability to conduct a defense of the airhead and to conduct a short-term sustainment of those forces. The intent of the buildup is to provide a secure operating and logistic base for forces working to move the airhead away from the original point of attack. Usually, this distance is equal to the enemy's direct fire capability to harass and destroy incoming aircraft or landing craft (5 to 10 kilometers).

4-77. Composing the follow-on echelon depends on METT-TC. All though not inclusive, it can consist of Armored, Stryker, Infantry, field artillery, air defense artillery, and combat engineers as well as other significant supporting elements (military information support operations, civil affairs, military police, and chemical, biological, radiological, and nuclear) and sustainment assets.

4-78. The time involved in the defense varies. It depends on the mission assigned, the composition and size of the force, the enemy reaction, and the type of operation contemplated. A well-prepared defense in short-duration missions in isolated objective areas may not be required. Security can be accomplished by completely or almost completely destroying or dispersing the enemy forces in the immediate objective area during the assault; then, airlifting the striking force before the execution of a coordinated enemy counterattack.

4-79. Defense of an airhead. The airborne force usually defends an airhead by securing key terrain within the airhead and dominating likely avenues of approach. Units deny the enemy the areas between the

Chapter 4

occupied positions with a combination of patrols, fires, and natural and man-made obstacles. Units aggressively reconnoiter between positions within the airhead and forward of the airhead line in the security. They increase emphasis on reconnaissance forward of the airhead line during limited visibility. The airhead configuration allows the commander to shift forces, reserves, and supporting fires quickly to reinforce other areas of the airhead. Regardless of the form of defense chosen, the force prepares positions in depth within its capabilities.

4-80. Defense during withdrawal. Should withdrawal from the initial positions be required, the final area to which the airborne force withdraws must contain adequate space for maneuver, for protecting critical installations, and for planned air landing or air evacuation operations.

4-81. Defense against armor. During the initial phases of an airborne operation, one of the main defenses against enemy armor is air support. Aircraft attack enemy armor targets as they appear, as far as possible from the objective area, and continue to attack and observe them as long as they threaten the airborne force. Strong points defending the airhead use existing obstacles such as ditches, thick foliage, and structures, plus reinforcing obstacles such as mine fields, tank traps, demolitions, and man-made obstacles. Units emplace AT weapons in depth along avenues of approach favorable for armor. They cover all dangerous avenues of approach with planned fires. Javelins, AT-4s and light anti-armor weapons of the rifle companies, the tube launched, optically tracked, wire guided weapons of the battalion weapons company, and the antitank weapons of division and corps aviation units give the airborne force a substantial amount of antitank firepower. Some of the antitank weapons, organic to battalions that are in holding areas not under armored attack, can be moved to reinforce threatened areas.

4-82. Defense against airborne assault: All personnel must recognize that the enemy can conduct airborne assault operations and must defend against these attacks. Helicopters afford the enemy one of its best means of rapidly moving significant operating forces to the airhead area.

4-83. Friendly forces must counterattack an enemy airhead immediately with available forces and fires to disrupt the enemy's plan and force build up, and continue until the enemy airborne assault has been neutralized.

4-84. Defense against guerrilla action and infiltration: The defense must include plans for countering enemy guerrilla attacks or infiltrated forces attacking the airhead area. The basic defense against these attacks is an extensive patrol and warning system, an all-round defense, and designated reserve units ready to move quickly to destroy the enemy force. Units must be especially alert during limited visibility to prevent the enemy from infiltrating. If the enemy can build up forces in the airhead interior, it can influence operations. Units must locate and destroy the enemy that has infiltrated the airhead.

SECTION V – FOLLOW-ON OPERATIONS

4-85. An airborne assault is as rapid in its execution as it is time-consuming in its preparation. Commanders must develop contingency plans for possible follow-on operations. These plans should be modified based on the most current intelligence. Advanced planning can allow more rapid decisionmaking and timely commitment of forces.

4-86. The employment of ABNAF on the ground is similar to that of other Infantry ground forces. The entire range of these operations include movement to contact, attack, area defense, or retrograde. Additional operations may include raids, linkup, relief in place, passage of lines, withdrawal (either overland or by air), exfiltration, and noncombatant evacuation. (Refer to FM 3-21.20, FM 3-21.10, and FM 3-21.8 for more information.)

SECTION VI – SUPPORTING OPERATIONS

4-87. Airborne forces can deploy from a continental United States base directly to the objective area. A more common method would be for the airborne unit to first deploy to a remote marshalling base or to an intermediate staging base (ISB) before establishing a lodgment in the area of operation. In certain circumstances, the objective can be beyond the range of aircraft operating from a remote marshalling base or ISB in friendly territory. Therefore, a forward operating base in hostile territory can be seized to facilitate or project further operations.

REMOTE MARSHALLING

4-88. The remote marshalling base is a secure base to which the entire airborne force (to include organic and attached support elements) deploys and continues mission planning. (See figure 4-4.)

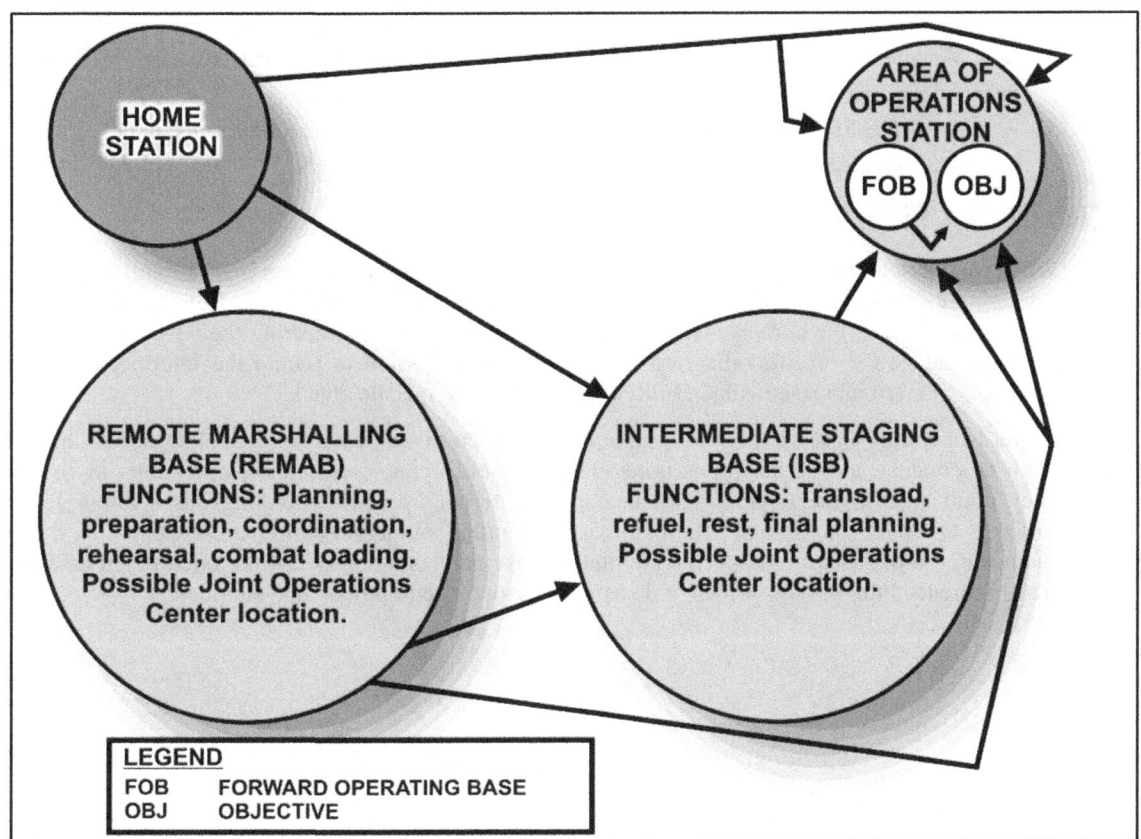

Figure 4-4. Base options

- Location. The remote marshalling base is within the geographical area encompassed by the command authority of the theater or joint task force commander. This ensures that the sustainment elements providing support to the airborne unit are operating within their normal area. It prevents or lessens out-of-area support requirements for sustainment elements. The remote marshalling base should be in an area similar in terrain and climate to the objective area. Time spent at the remote marshalling base lets the unit begin acclimatization.
- Planning and coordination. The remote marshalling base provides a secure location for the unit to conduct detailed planning and coordination with the controlling headquarters staff.
- Preparation. In the remote marshalling base, the commander conducts rehearsals, refines and modifies plans, determines priority intelligence requirements, and coordinates with the proper intelligence source to receive that information.
- Additions to the unit. In the remote marshalling base, individual specialists who augment the force are integrated into the unit if they have not already joined. Specially trained supporting units (such as aviation and communication elements) join the force at the remote marshalling base.
- Functions of a remote marshalling base. The remote marshalling base must provide—
  - Access to the controlling headquarters staff.
  - Physical security of billeting, planning, maintenance, and communication areas.

Chapter 4

- Mess, billeting, latrine, and shower facilities for the force and its supporting elements.
- Access to a C-17 or C-130 capable airfield, possibly with all-weather operations.
- Access to secure communication and processed intelligence.
- Access to rehearsal areas where sites can be built and live-fire rehearsals can be conducted.
- Access to the unit locations of major supporting elements such as naval landing craft or Army aviation units.
- An external security force and an active counterintelligence agency.
- Vehicle transport for personnel lift, equipment transfer, and administrative use.
- Access to maintenance support facilities.
- Army Health System support facilities to augment the airborne medical personnel.
- Covered areas for packing parachutes and rigging airdrop loads.

INTERMEDIATE STAGING BASE

4-89. When the assaulting force must move a considerable distance from its sustaining base, the commander should consider establishing one or more intermediate staging bases. Before establishing an intermediate staging base, the commander must weigh the benefits of establishing the base against the cost in terms of combat power or effort diverted from the support mission to secure the intermediate staging base, and the potential of increased vulnerability to enemy air and missile attack.

4-90. Forward operating bases extend and maintain the operational reach by providing secure locations from which to conduct and sustain operations. They not only enable extending operations in time and space; they contribute to the overall endurance of the force. Forward operating bases allow forward deployed forces to reduce operational risk, maintain momentum, and avoid culmination. Generally, they are located adjacent to a distribution hub. This facilitates movement into and out of the operational area while providing a secure location through which to distribute personnel, equipment, and supplies.

Chapter 5

Landing Plan

The landing plan supports the ground tactical plan. It provides a sequence for the arrival of forces into the area of operation, ensuring that all forces arrive at designated locations and times prepared to execute the ground tactical plan. The commander finalizes the landing plan after completing the ground tactical plan. The landing plan phases forces into the objective area at the correct time and place to execute the ground tactical plan. Executing the landing plan is vital to the swift massing of combat power, protecting the force, and subsequent mission accomplishment.

SECTION I – DELIVERY CONSIDERATIONS

5-1. The landing plan is the ABNAFC's plan that links the air movement plan to the ground tactical plan. It is published at ABN IBCT level and below. Before the ABNAFC can prepare an overall landing plan, he must know where his subordinate commanders intend to land their assault forces. The landing plan is generated up the chain of command as a collaborative effort.

ORGANIZATION

5-2. Airborne forces organize landing plans to maintain tactical integrity. Battalions or battalion task forces normally land intact on a single drop zone. Two or more battalions land successively on the same drop zone or each can land on a separate drop zone within a general ABN IBCT area of operation. The ABN IBCT lands in mutually supporting drop zones.

5-3. The ABNAF sends as many assault force personnel and equipment as possible into the area in airborne assault serials. The ABNAFC must consider the mobility of equipment after the landing. For example, the carriers or prime movers that are deliverable by parachute, but difficult to manhandle on the ground, can accompany the weapons in the assault force. Assault forces accompany their units' principal items of equipment.

REQUIREMENTS

5-4. To develop the landing plan, commanders at each level need to know their commander's priorities, the airlift plan, the landing area study, the parent and subordinate unit task organization and ground tactical plans, and subordinate unit landing plans. During the backbrief of the ground tactical plan, the commander establishes airlift and delivery priorities and airlift plan. He provides as much of this information as possible to subordinate units at the end of the ground tactical plan backbrief.

5-5. As with the ground tactical plan, each echelon (fire team through brigade) must conduct confirmation briefs and backbriefs to their landing plans. The landing plan remains tentative until leaders complete backbriefs and coordinate changes. The leaders take the following actions:

- Landing plan. In the case of the landing plan, backbriefs ensure coordination of who is using what drop zone, and or landing zone and when, the preferred orientation of drop zones, and who is landing in which areas and when. The landing plans follow the commander's priorities, the airlift plan, and ground tactical plan.
- Assembly plan. One of the most critical parts of the landing plan is the assembly plan. Each leader must brief his Soldiers, require a backbrief, rebrief his Soldiers, and require another backbrief. Each Soldier should know exactly what to do, how to do it, and when to do it to assemble quickly. Assembly plans of one unit do not interfere with the assembly plans of other units.

Chapter 5

- Aircraft requirements. The backbrief of the landing plan identifies aircraft requirements for each subordinate unit. If there are not enough aircraft available to lift the entire assault force at one time, commanders must decide the units that should be lifted first, and then allocate aircraft accordingly. In making this decision, he analyzes the priorities dictated by the mission and the higher commander.

PRIORITIES

5-6. Commander must set the priorities for each assault objective to determine the delivery sequence for units that are to secure these objectives. This does not necessarily match the sequence in which the units secure objectives. The commander must know—

- Priorities for deliveries on each drop zone (personnel drop, container delivery system, and heavy drops).
- The drop zone sequence.
- Priorities for delivering the remainder of the forces.
- Method of delivery for each unit and its equipment.
- The location of the heavy equipment point of impact, and the personnel point of impact.
- Abort criteria (Go/No-Go) and bump plan.

5-7. Airlift plan. The USAF airlift planners responsible for the airlift plan develops the plan, in coordination with ABNTF planners, to best support the ground tactical plan. This plan includes aircraft formations and the sequence of personnel drops, and heavy drops. ABNTF planners choose the sequence and the time intervals between serials, which are groups of like aircraft (C-130s, C-17s) with the same delivery method (personnel drops, heavy drops,) going to the same drop zone.

5-8. Landing area study. Division or corps staff, working with U.S. Army topographic engineers and the USAF, develops the landing area study and provides it to subordinate units. This study enables subordinate units to select the location, size, and orientation of drop zones to best support their scheme of maneuver.

5-9. Subordinate unit landing plans. Subordinate commanders should develop landing plans to support their own respective schemes of maneuver. Subordinate units then backbrief their landing plans so that higher headquarters can finalize their plans. Units must know the initial locations of sustainment assets. This information should become available as subordinate units backbrief their ground tactical plan.

SECTION II – DELIVERY ELEMENTS

5-10. A designated lodgment area in a hostile or potentially hostile territory, when seized and held, makes the continuous landing of troops and materiel possible and provides maneuver space for the transition to follow-on operations. The five elements of the landing plan include: sequence of delivery, method of delivery, place of delivery, time of delivery, and assembly plan. Assembly and reorganization are discussed in Section IV of this chapter.

SEQUENCE OF DELIVERY

5-11. The ABNAFC's priorities within the ground tactical plan determine the sequence of delivery, with the initial assault designed to surprise and concentrate overwhelming combat power against the enemy and to protect the assault force. Neither aircraft allocations nor the availability of aircraft should influence these decisions. He determines final aircraft allocations after the landing plan confirmation brief.

5-12. Advance serials may precede the main airlift column to drop USAF combat control teams and Army long-range surveillance teams. The combat control team places and operates navigation aids on the drop zones and landing zones; the long-range surveillance teams provide surveillance on named areas of interest and reports to the ground force commander. Insertion location of advance serials whether within the objective area or some distance away from the objective area are always METT-TC dependent, as the commander strives to achieve surprise regarding exact objectives within the airborne operations.

METHOD OF DELIVERY

5-13. This part of the landing plan addresses how the force with its needed supplies and equipment arrives in the objective area. The assault echelon comes in by parachute. The ABNAFC can use a number of other means to introduce additional personnel, equipment, and supplies into the objective area.

PERSONNEL AIRDROP

5-14. The airborne force delivers assault personnel by parachute drop. This method allows quick, nearly simultaneous delivery of the force. Planners choose terrain with minimal obstacles that allows the assault force to land on or close to objectives. In some cases and with special equipment, it can deliver personnel into rough terrain. Special teams can use high-altitude high-opening (HAHO) or high-altitude low-opening (HALO) parachute techniques. These methods allow for early delivery without compromising the objective's location.

EQUIPMENT/SUPPLY AIRDROP

5-15. Airborne forces can airdrop supplies and equipment directly to ground forces behind enemy lines or in other unreachable areas. However, there are advantages and disadvantages.
- Advantages are as follows:
  - Prerigging and storing emergency items for contingencies considerably reduces shipping and handling time and increases responsiveness.
  - Since the delivery aircraft does not land, there is no need for forward airfields, landing zones or material handling equipment for offloading.
  - This reduces flight time and exposure to hostile fire and increases aircraft survivability and availability.
  - Ground forces can disperse more since they are not tied to an airfield or strip.
- Disadvantages are as follows:
  - Airdrops require specially trained rigger personnel and appropriate aircraft.
  - Bad weather or high winds can delay the airdrop or scatter the dropped cargo.
  - Ground fire threatens vulnerable aircraft making their final approach, especially if mountains or high hills canalize the aircraft.
  - Since the aircraft do not land, no opportunity for ground refueling exists. Planned aerial refueling can extend aircraft range and should be considered on long flight legs to increase objective area loiter time and mission flexibility.
  - Bulky airdrop rigs for equipment prevent the aircraft from carrying as much cargo as when configured for airland.
  - The possibility of loss or damage to equipment during the airdrop always exists.
  - Ground forces must secure the drop zone to prevent items from falling into enemy hands.
  - Recovery of airdropped equipment is slow and manpower intensive.

Types of Delivery

5-16. All means of delivery are exploited to maximize combat power in the lodgment. Free drop, high-velocity airdrop, low-velocity airdrop, and joint precision airdrop, are different types of airdrop delivery.

Free Drop

5-17. Less than 600 feet above ground level (AGL), free drop requires no parachute or retarding device. The airdrop crew can use energy-dissipating materiel around the load to lessen the shock when it hits the ground at a rate of 130 to 150 feet a second. Fortification or barrier material, clothing in bales, and other such items can be free dropped.

Chapter 5

High-Velocity Airdrop

5-18. Parachutes, which have enough drag to hold the load upright during the descent at 70 to 90 feet a second, stabilize loads for high-velocity airdrops. Army parachute riggers place airdrop cargo on energy-dissipating material and rig it in an airdrop container. This method works well for subsistence, packaged petroleum, oils, and lubricants (POL) products, ammunition, and other such items. The ground commander may use the standard high-velocity delivery system, which is the container delivery system, to deliver accompanying and follow-on supplies; they can be delivered within an area 400 by 100 meters. A container delivery system is the most favored means of resupply; it is the most accurate of all airdrop methods. Each pallet holds up to 2200 pounds. A C-130J holds up to 16 of these containers, and a C-17 holds up to 40 of these containers. Planners should calculate the computed air release point (CARP) near assembly areas or resupply points. The air liaison officer or Army drop zone support team controls receipt of container delivery systems.

Low-Velocity Airdrop

5-19. 700 ft AGL to 1300 ft AGL. Low-velocity airdrop requires cargo parachutes. Crews rig items on an airdrop platform or in an airdrop container. They put energy-dissipating material beneath the load to lessen the shock when it hits the ground. Cargo parachutes attached to the load reduce the rate of descent to no more than 28 feet a second. Fragile materiel, vehicles, and artillery can be low-velocity airdropped.

- Heavy drop. Airborne forces use this method most often to deliver vehicles, bulk cargo, and equipment. Airdrop aircraft deliver heavy-drop equipment just ahead of the main body or, if following personnel drops, at least 30 minutes after the last paratrooper exits. For night drops, the heavy-drop precedes personnel drops.
- Door bundles. Requires the use of either the A7A cargo sling or the A21 cargo bag. With these, aircraft personnel can drop unit loads of up to 500 pounds just before the first Soldier's exit. Airdrop standard operating procedures dictate the number and type of door bundles that specific aircraft can drop.

Joint Precision Airdrop

5-20. The Joint Precision Airdrop System (JPADS) is a high altitude-capable guided precision airdrop system that provides increased control release from the aircraft, and reduces on-ground load dispersion with GPS-supported accuracy. Two current increments of JPADS support precision delivery of cargo pallets up to 2200 pounds and up to 10,000 pounds.

Airland

5-21. Airborne forces can accomplish certain phases of airborne operations, or even the entire operation, by using airland to deliver personnel and equipment to the objective area. The advantages and disadvantages are as follows:

- Advantages. In some cases, air landing rather than air-dropping personnel and equipment may be advantageous because air landing—
  - Provides the most economical means of airlift.
  - Delivers Army aviation elements, engineering equipment, artillery pieces, and other mission-essential items in one operation.
  - Provides a readily available means of casualty evacuation.
  - Allows forces to more easily maintain tactical integrity and to deploy rapidly after landing.
  - Allows the use of forces with little special training and equipment.
  - Does not require extensive preparation and rigging of equipment.
  - Offers a relatively reliable means of personnel and equipment delivery regardless of weather.
  - Precludes equipment damage and personnel injuries forces may experience in parachute operations.
- Disadvantages. In other cases, air landing is not advantageous because it—
  - Cannot be used for forced entry.

- Requires moderately level, unobstructed landing zones with adequate soil trafficability.
- Requires more time for delivery of a given size force than airdrop, especially for small, restricted landing zones.
- Generally requires improvement or new construction of airland facilities, which adds to the engineer workload.
- Requires some form of airlift control element support at offloading airfields. Mission intervals depend on airlift control element size, offloaded equipment availability, and airfield support capability.

Airland Organization for Movement

5-22. The tactical integrity of participating forces is a major consideration in an airland operation. Small forces that are expected to engage in combat on landing, airland organizationally intact with weapons, ammunition, and personnel in the same aircraft. Airland planning stresses placing forces as close as possible to objectives, consistent with the availability of landing zones and the operational capability of the aircraft employed. Because of aircraft vulnerability on the ground, forces unload as quickly as possible.

5-23. The airborne commander determines the composition of each aircraft load and the sequence of delivery. The mission, the tactical situation, and the assault force's task organization influence this decision.

5-24. Forces should use facilities, such as roads and open areas, to reduce the time and effort required for new construction. They should consider layouts that facilitate future expansion and provide maximum deployment and flexibility. As the size and efficiency of an air facility improve, its value to the enemy as a target increases. To reduce this vulnerability, the facilities should be dispersed and simple.

PLACE OF DELIVERY

5-25. Selecting drop zones and landing zones is a joint responsibility. The airlift commander is responsible for the precise delivery of personnel and cargo to the drop zone or landing zone and selecting approaches to the drop zone. Both joint and component commanders must base their decisions on knowledge of their respective problems and on the needs of the overall operation. The nature and location of landing areas are important considerations when preparing the scheme of maneuver. The general area in which they are to be established is governed by the mission. At higher echelons, commanders can assign landing areas in broad general terms. In subordinate units, leaders must describe their locations more specifically. Drop zones are selected after a detailed analysis. If the enemy situation permits, the commander should choose a drop zone directly on top of assault objectives. Commanders should consider the following factors when making their selections:

- Ease of identification. The drop zone should be easy to spot from the air. Airlift pilots and navigators prefer to rely on visual recognition of terrain features to deliver personnel and equipment in an accurate manner.
- Straight-line approach. To ensure an accurate airdrop, the aircraft makes a straight-line approach to each drop zone for at least 10 miles, or about four minutes at drop airspeed, before the start of the drop.
- Out of range. The commander should choose a drop zone that allows the forces to avoid enemy air defenses and strong ground defenses and puts them outside the range of enemy suppressive fires. To get to the drop zone, aircraft should not have to fly over or near enemy antiaircraft installations, which can detect aircraft at drop altitudes. They should fly over hostile territory or positions for the least possible time.

5-26. Suitable weather and terrain. The commander must consider the weather and terrain due to conditions that may affect the usability of a drop zone.

- Weather. Commanders should consider seasonal weather conditions when selecting drop zones. Adverse weather effects can be devastating. Ground fog, mist, haze, smoke, and low-hanging cloud conditions can interfere with the aircrew's observation of drop zone visual signals and markings. However, they do offer excellent cover for blind or area drop zones. Excessive winds hamper operations.

- Terrain. Flat or rolling terrain is desirable; it should be relatively free, but not necessarily clear, of obstacles. Obstacles on a drop zone do not prevent paratroopers from landing but increase jump casualties. Sites in mountainous or hilly country with large valleys or level plateaus can be used for security reasons. Small valleys or pockets completely surrounded by hills are difficult to locate and should be used only in rare cases. Commanders must avoid man-made obstacles more than 150 feet (46 meters) above the level of the drop zone within a radius of three nautical miles. High ground or hills need not be considered a hazard unless the hills pose an escape problem that is beyond the aircraft's capability. High ground or hills more than 1000 feet (305 meters) above the surface of the site should not be closer than three nautical miles to the drop zone for night operations. The perimeter of the drop zone should have one or more open approach areas free of obstacles that would prevent the aircrew's sighting of the drop zone markings.
  - Cover and concealment. Cover and concealment near the drop zones or landing zones are a distinct advantage when the airborne forces assemble and when airland forces land.
  - Road network. Having a drop zone near a good road network expedites moving personnel, supplies, and equipment from that zone. If the landing area contains terrain that is to be developed into an air landing facility, a road network is of particular value not only for moving items from the facility but also for evacuating personnel and equipment as well. However, a road network offers the enemy a high-speed avenue of approach and must be defended.
  - Key terrain. The drop zone site selected should aid in the success of the mission by taking advantage of dominating terrain, covered routes of approach to the objective, and terrain favorable for defense against armored attack.

5-27. Minimum construction for drop zones or landing zones. Because of limited engineer support in the airborne force, selected landing zones should have a minimum requirement for construction and maintenance. Unless more engineer support is requested and received, construction and maintenance restraints can limit the number of areas that can be used or developed.

5-28. Mutual support. Commanders should select mutually supporting drop zones or landing zones that provide initial positions favorable to the attack.

5-29. Configuration. The division or brigade commander gives guidance on drop size in operations plans or operations orders. Then each unit commander determines the exact shape, size, and capacity required.
- Shape. The most desirable shapes for drop zones are rectangles or circles; these permit a wider choice of aircraft approach directions. However, they require precise navigation and timing to avoid collisions or drop interference.
- Size. The drop zone should be large enough to accommodate the airborne force employed; one drop zone that allows the aircraft to drop its entire load in one pass is desirable. Repeated passes are dangerous because the initial pass can alert enemy antiaircraft and other emplacements, waiting for subsequent drops.
  - There are certain situations, however, when multiple passes are used. This occurs mainly when there is no significant air defense threat and orbits are made over areas where enemy antiaircraft systems are not positioned. This applies especially to the seizure of islands where small drop zones are the rule. If there are enough aircraft available to deliver the force with fewer personnel on each aircraft, there is no real problem. However, if there are only enough aircraft to deliver the assault echelon in one lift with each aircraft carrying the maximum number of personnel, then the aircraft has to make multiple passes over the drop zone.
  - A large drop zone can permit several points of impact to be designated and used. It is desirable to saturate the objective area in the shortest possible time but there is a reasonable limit to the amount of personnel and heavy equipment drop that can be stacked on a single drop zone. Therefore, it can be desirable to use multiple points of impact on a single drop zone provided the drop zone is large enough to permit this.
- Capacity. The drop zone capacity is based on the expected number of forces to be dropped and their dispersion pattern.

Landing Plan

5-30. Orientation. Thoughtful orientation allows the quickest possible delivery of the airborne force into the objective area.
- Ideal drop zones offset and parallel each serial. (See figure 5-1.) This allows aircraft to share a flight route until they approach the objective area; then they can split at an initial point or release point for simultaneous delivery on several drop zones.
- Another method is parallel on-line. (See figure 5-2, page 5-8.) This employs making two drops on two drop zones in line , which eliminates a change of flight direction between the two drops. The drop zones must be far enough apart to permit the navigators to compute the location of the second release point.
- Paratroopers are more likely to overshoot the drop zone than to undershoot it. The primary objective for selecting the trailing edge drop zone for the primary assault is for personnel in the front of the aircraft to exit last.
- If a fighter aircraft escort or rendezvous is required for the drop, they must remain advised of the drop pattern, the direction of all turns to be flown around the drop zone, and the areas to look for possible enemy activity. Drop zones that require intersecting air traffic patterns should be avoided, whenever possible. They delay simultaneous delivery of the force because of the safety requirements to stagger delivery times and clear the air by at least a 5- or 10-minute formation separation time. They require that joint suppression of enemy air defenses (SEAD) is accomplished for multiple routes instead of one. This may result in piecemeal delivery and an unnecessarily complicated plan, violating the principles of mass and simplicity.

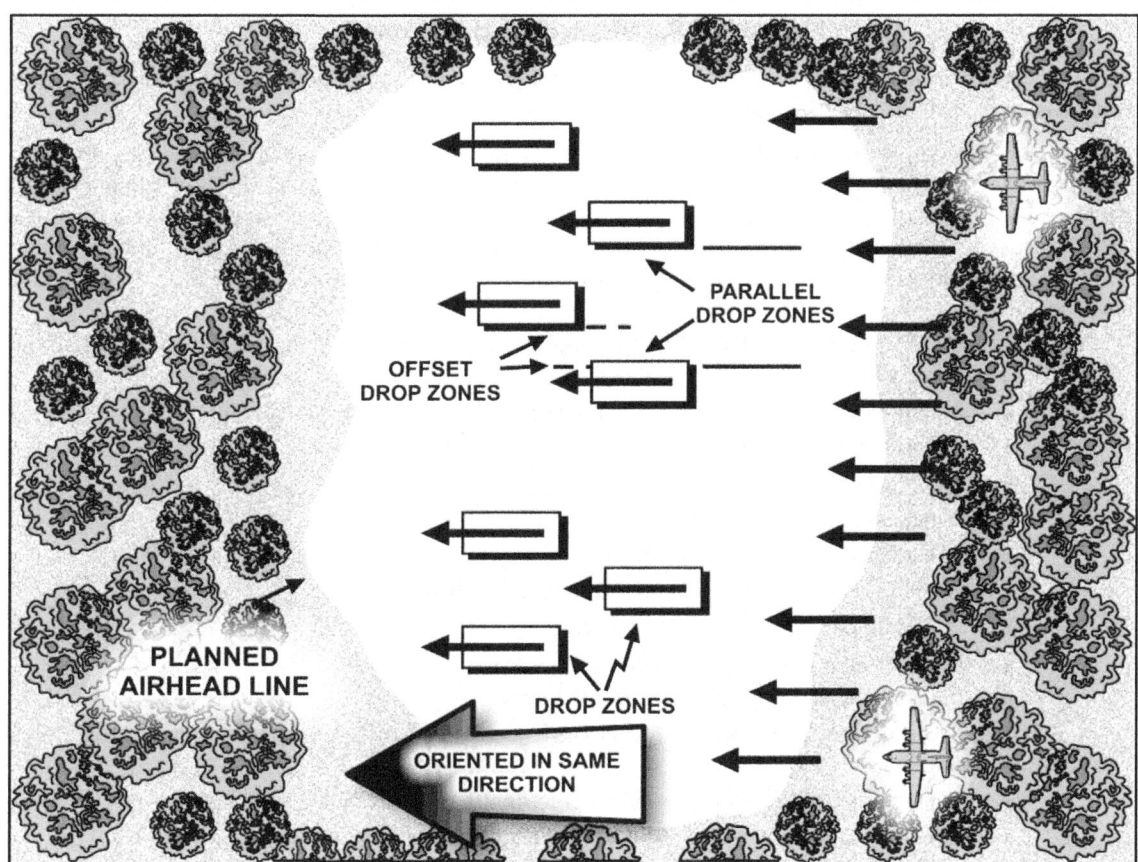

Figure 5-1. Offset and parallel drop zones

Chapter 5

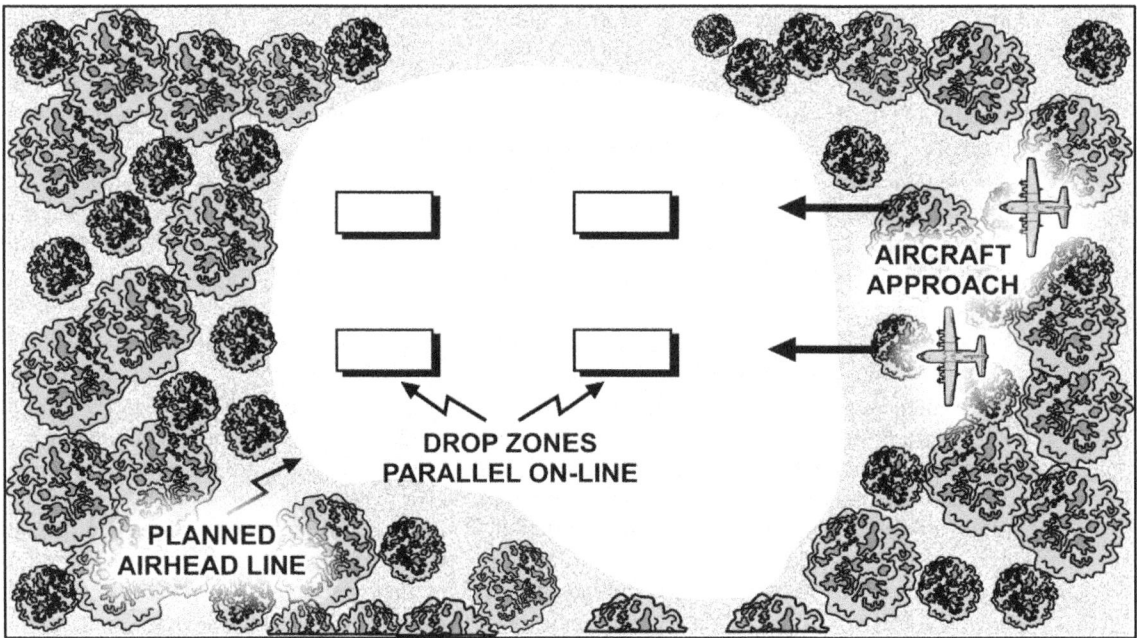

Figure 5-2. Parallel on-line drop zones

5-31. **Alternate drop zones or landing zones.** Commanders must select alternate drop zones or landing zones to compensate for changes that may occur.

5-32. **Number of drop zones or landing zones.** The number of drop zones to be used by the assault parachute echelon of an ABN IBCT depends on the number, size, and relative position of suitable sites; the ABN IBCT plan of maneuver; and the expected enemy situation. The maneuver battalions of an ABN IBCT can land successively on the same drop zone, on separate battalion drop zones, or on adjacent areas within a single large ABN IBCT drop zone. Advantages and disadvantages of each drop zone is address below:
- Single BCT drop zone. The use of a single ABN IBCT drop zone on which battalions land successively has the following advantages:
    - It permits greater flexibility in the plan of maneuver and the plan of supporting fires.
    - It facilitates coordinating and controlling assault battalions.
    - It applies the principle of mass.
    - It makes logistical support easier.
    - It decreases the area of vulnerability.
- The use of a single drop zone has the following disadvantages:
    - It slows the buildup of combat power.
    - It causes later airlift sorties to be vulnerable to enemy air because of the loss of surprise.
    - It allows the enemy to emphasize his efforts.
- Separate battalion drop zones. The use of separate battalion drop zones has the following advantages:
    - It increases readiness for action by deploying the ABN IBCT as it lands.
    - It reduces confusion on the drop zones during the landing and reorganizing.
    - It tends to deceive the enemy as to the intention and strength of the landing force.
    - It makes capture of the ABN IBCT objective easier when there is strong opposition on one drop zone.
    - It increases the freedom of maneuver of the assault battalions.
- The use of separate battalion drop zones has the following disadvantages:
    - It makes mission command more difficult.

- It reduces flexibility because forces are dispersed.
- Adjacent drop areas. Landing battalions on adjacent areas within a single large ABN IBCT drop zone has, although to a lesser degree, the same advantages and disadvantages of dropping on separate drop zones.

TIME OF DELIVERY

5-33. No set rule can be prescribed for the timing of an airborne operation. It varies with each situation; however, the airborne force tries to conduct airborne assaults during limited visibility to protect the force and to surprise the enemy. The commander sets the specific time of delivery. However, for the landing plan, times are stated in terms of P-hour. *P-hour* is the specific hour on D-day at which a parachute assault commences with the exit of the first Soldier from an aircraft over a designated drop zone. P-hour may or may not coincide with H-hour (FM 6-0). The following considerations affect the timing of the operation.

SUPPORT OF THE DECISIVE OPERATION

5-34. The airborne assault can be a shaping operation. If so, the time of commitment of the airborne forces in relation to the decisive operation is directed by orders from higher headquarters. It is determined in advance according to the mission, the situation, and the terrain. For example, the airborne force can be committed in advance of the decisive operation to give the airborne assault an increased element of surprise. It can be committed during the decisive operation to neutralize specific areas or to block the movement of enemy reserves. It can be committed after the decisive operation to assist a breakthrough or to block an enemy withdrawal.

VISIBILITY

5-35. The decision as to whether the airborne force is committed by night or day depends on the estimated degree of air superiority, the need for security from enemy ground observation, the relative advantage to be gained by surprise, and the experience of both airlift and airborne personnel.

5-36. As an advantage, night airborne operations greatly increase the chance of surprise and survivability, and reduce the chance of attack by enemy aircraft during the air movement. They reduce vulnerability to antiaircraft fire, conceal preparations for takeoff from the enemy, and reduce the effectiveness of the defender's fires. Daylight operations provide better visibility from the air and ground, more accurate delivery, quicker assembly, and more effective friendly fires than night operations.

5-37. As a disadvantage, night airborne operations in zero visibility require well-trained Soldiers and aircrews to locate the drop zone and assemble rapidly. They provide more air and land navigation problems and offer slower rates of assembly than day operations. Night operations reduce the effectiveness of close air support. Day operations increase vulnerability to enemy air defense, ground fires, and air attack, and they result in loss of surprise.

INTERVALS

5-38. The time interval between delivery of the assault echelon (P-hour) and the follow-on echelon depends on the availability of aircraft, the capacity of departure airfields, the number of aircraft sorties that can be flown on the initial airborne assault, the availability of drop zones or landing zones within the objective area, and the enemy situation. For example; if there are unlimited aircraft, ample departure airfields, numerous drop zones or landing zones within the objective area, and little or no enemy air defense, the commander could deliver the follow-on echelon immediately after the assault echelon. Thus, the time interval could be so brief that it would be hard to determine which was the last aircraft of the assault echelon and which was the first aircraft of the follow-on echelon. Regardless of the timing selected, avoid setting a pattern.

SECTION III – PREPARATION AND SUPPORTING FIRES

5-39. An ABNAF's organic fire support is typically limited to its mortars when initially conducting the airborne assault phase of an entry operation. The ABNAFC must plan for the use of fire support external to

Chapter 5

the ABNAF until organic field artillery assets are delivered by heavy-drop or with the airland phase of the operation. Once paratroopers de-rig and prepare their gun systems from the heavy-drop or airland and place into action those systems, the ABNAFC then can use their supporting fires. As part of the landing plan and following the ground tactical plan, the ABNAFC and staff must plan and integrate a fire support plan as part of the operation.

FIRE SUPPORT PLANNING

5-40. Fire support planning for an airborne operation is initiated on receipt of the mission. Concurrent with the development of the concept of the operation, the commander plans for fire support so that it is provided throughout the operation.

PLANNING, COORDINATION, AND EXECUTION

5-41. Fire support teams, elements, and cells advise the commander on fire support capabilities and joint fire support command and control, effective use of fires assets, and assist in the planning, coordination, and execution of fires. Fire support planning, coordination, and execution for airborne operations are more complex than ground operations not involving vertical envelopment due to the following factors.

5-42. The assault elements of the airborne force are quickly placed in direct contact with the enemy deep in hostile territory. Initial operations are decentralized and communications can be limited or nonexistent. During the initial airborne assault and periodically thereafter, airspace over the drop zone contains a high density of airdrop aircraft, which complicates fire support aspects of airspace management.

5-43. Airborne force vulnerability increases during the time between landing and assembly or seizure of assault objectives. This time varies based on force size and the mission variables of METT-TC. During this vulnerable period, reliable communications are essential to coordinating and executing fire support missions.

5-44. Calls for fire are sent under conditions where forces are in critical need of fire support. Units lack firm knowledge of the situation, especially locating friendly and enemy forces. Calls for fire can come when reliable ground communications have not been firmly established.

5-45. Initially, artillery support in the airhead is limited. This situation occurs at the same time as the arrival of the assault echelon or the operation's opening phase. Consequently, the bulk of fire support must come from joint fire support assets and organic mortars. Fire support can be provided by the long-range artillery of advancing friendly forces (if in range), Army long-range rocket or missile fire, and joint interdiction.

INITIAL AVAILABLE FIRE SUPPORT

5-46. Joint fire support, organic mortars, and limited Army aviation and field artillery are usually the only fire support available to the airborne force until the lodgment is established. Examples of initial available fire support missions include:
- Column cover for the assault and follow-up echelons and resupply sorties.
- Suppression of enemy air defenses along the corridor selected for penetration and near the objective.
- Counterair operations to gain and maintain air superiority along the corridor and in the objective area.
- Pre-assault fires of the airhead and other critical targets and deception.
- Field artillery may conduct counterair missions by striking enemy aircraft bases, helicopter forward aiming and refuel points, missile storage, and launch sites.
- Air interdiction of the objective area to include armed reconnaissance missions targeted against enemy forces that react to the airborne assault.
- Air defense of marshalling areas, resupply airfields, and the airhead.
- Close air support.

FIRE SUPPORT ASSETS

5-47. Fire support assets can perform a variety of missions in support of the airborne assault. The following are examples of standard missions arranged by type of asset.

Joint Air Support

5-48. Joint air support, preplanned and immediate, provides air interdiction and close air support to an airborne operation. Types of support include:
- Air interdiction pre-assault fires within the objective area and other critical targets.
- Suppression of enemy air defenses.
- Electronic warfare.
- Close air support to the airborne assault force.

Naval Gun Fire

5-49. Naval gun fire, when available and in range, is a reliable, accurate, high-volume source of fire support. Naval gun fire provides the following types of support:
- Pre-assault fires of the objective area and other critical targets.
- Suppression of enemy air defenses.
- Direct support and general support of forces in contact.
- Interdiction (land and sea).

Field Artillery

5-50. Artillery of linkup forces within range can provide the following support:
- Interdiction fires.
- Suppression of enemy air defense fires.
- Counterfire.
- Fires to maneuver units.

Army Aviation

5-51. Army aviation can conduct close combat attack when the intermediate staging base or forward staging base is within range or when a secure airfield permits airland and buildup of Army aviation that is transported in USAF airlift aircraft. They can support—
- Interdiction of enemy reaction forces, especially mechanized forces with accurate, long-range antitank fires.
- Seizure of assault objectives with rocket fire and gunfire.
- Reconnaissance and surveillance tasks within and beyond the security area.

FIRE SUPPORT CONSIDERATIONS

5-52. Fire support planning relies on careful, thorough planning based on fire support principles designed to support maneuver. Fire support considerations include the following.

UNITY OF CONTROL AND CONTINUOUS LIAISON

5-53. Unity of control may be met through the establishment of joint headquarters (such as a joint task force) to include a joint operations center or other higher headquarters, which is responsible for coordinating fire support to the maneuver commander. Liaison, especially between Army and USAF units, is necessary at all echelons down to battalion/squadron level. It must be supported with adequate communications to facilitate command and to control lateral dissemination of information and coordination. Joint agreements, memorandums of understanding, standard operating procedures, and signal operating instructions all facilitate this effort. Example, each assault battalion and ABN IBCT must have attached tactical air control party (TACP) and naval gun fire line of sight, if naval gun fire is available.

Chapter 5

CENTRALIZED COORDINATION

5-54. Due to the nature of the airhead and the required continuous airflow into the airhead, fire support assets must be closely controlled to prevent fratricide and waste of assets. During the initial stages of an airborne operation and before adequate ground communications can be established, coordination and control of fire support are accomplished from an airborne platform.

5-55. On landing, each battalion/squadron or BCT or regimental headquarters establishes contact with the airborne platform or joint airborne communications center/command post (JACC/CP) through the TACP. Fire support, such as close air support, beyond that available from organic or direct support assets would be requested from the airborne platform. Prioritizing and coordinating requests are accomplished by the ground force commander's representative in the airborne platform. His responsibilities include the following:
- Prevent fratricide of ground personnel.
- Ensure that requests do not interfere with incoming serials, other aircraft, or naval operations.
- Determine the fire support means to be employed in coordination with appropriate commander's staff members.
- Determine, while coordinating with the command staff, added safety or control measures required then transmit them to the appropriate ground elements.

5-56. For air missions, the commander's staff establishes contact with the appropriate flight, provides essential information, and then hands the flight off to the appropriate TACP or forward air controller for mission execution. At that point, the mission is conducted the same way as conventional operations. If naval gun fire or air support is available, it is essential that a naval gunfire liaison officer be present in the airborne platform to perform a similar function.

5-57. Once adequate facilities have been established in the airhead, fire support coordination responsibilities are passed from the airborne platform to the ground commander and his staff. (There is no doctrinal time for this transfer.) In some situations (for example, raids), this cannot occur; however, once an ABN IBCT main or tactical command post is on the ground transfer takes place in most cases.

COORDINATION MEASURES

5-58. The ABNAFC, assigned an area of operation within the airhead, employs fire support coordination measures (FSCMs) to facilitate rapid target engagement while simultaneously providing safeguards for friendly forces. Boundaries are the basic FSCM. The fire support coordinator recommends additional FSCMs to the commander based on the commander's guidance, location of friendly forces, scheme of maneuver, and anticipated enemy actions. FSCMs are either permissive or restrictive.

5-59. The primary purpose of permissive measures is to facilitate the attack of targets. Once they are established, further coordination is not required to engage targets affected by the measures. Permissive FSCMs include a coordinated fire line (CFL), a fire support coordination line (FSCL), and a free-fire area (FFA).

5-60. The primary purpose of restrictive measures is to provide safeguards for friendly forces. A restrictive FSCM prevents fires into or beyond the control measure without detailed coordination. Restrictive FSCM include an airspace coordination area, a no-fire area (NFA), a restrictive fire area (RFA), and a restrictive fire line (RFL). Establishing a restrictive measure imposes certain requirements for specific coordination before the engagement of those targets affected by the measure. (Refer to FM 3-90-1 for a detailed discussion of FSCMs.)

AIR MOVEMENT PLAN

5-61. Fire support during movement to the objective area is the responsibility of the airlift commander and staff. However, the airborne force commander must be closely involved because of the possibility of downed aircraft or a mission being diverted. Planning considerations include—
- Ensure fire support personnel and equipment is included on load plans and manifests.
- Plan targets on enemy.

Landing Plan

ARTILLERY AND MORTAR EMPLOYMENT

5-62. The initial phase of the airborne operation is decentralized and flexible until the assault objectives are secured and the airhead is established. During parachute assaults, the organic field artillery battalion of the ABN IBCT provides field artillery support within the context of assigned tactical missions. Airborne artillery adheres to tactics and methods applicable to other artillery units. Mortars provide rapid fire support at the company and battalion level. If the battalion commander chooses to employ his mortars as a split section, he also must plan how the mortar fire direction center will control fires.

NAVAL GUNFIRE

5-63. When operating on islands or near a coastline, naval gun fire support may be available to the airborne force. Naval guns can provide high-volume, accurate fires employing a variety of ammunition.

Air and Naval Gunfire Liaison Company

5-64. The air-naval gunfire liaison company (ANGLICO) provides ship-to-shore communications and long-range, fire control teams to adjust fire. In the absence of ANGLICO fire control teams, the fire support team can call for and adjust fires through the ANGLICO team.

ANGLICO Organization

5-65. Deployed ANGLICO forces comprise a command element, operational element (air/naval gun fire teams), and support element. Forces are as follows:

- The company is organized into groupings. The headquarters or support section and divisional air/naval gun fire section give command, control, administration, training, and logistics support for the company. They provide fire support planning and liaison personnel to the airborne force. Three ABN IBCT air/naval gun fire platoons provide liaison and control for air and naval gun fire to the assault companies, battalion, and the ABN IBCT.
- Each ABN IBCT platoon divides into an ABN IBCT team and two supporting arms liaison teams, which support two forward battalions. Each supporting arms liaison team has two firepower control teams, which support the forward companies of the battalions.
- The ANGLICO assists the staff in matters concerning air and naval gun fire. It coordinates requests for air and naval gun fire support from the battalions and squadron of the ABN IBCT and represents the airborne mission control platform, if required.
- The liaison officer and firepower control teams operate in the ground spot network. They communicate with the ship by HF radio to request and adjust naval gun fire. The firepower control team communicates with the liaison officer, using VHF radios. The liaison officer can communicate with aircraft using UHF radios.

Tactical Missions

5-66. Naval gunfire ships are assigned one of two tactical missions—direct support or general support. Missions are as follows:

- Direct support. A ship in direct support of a specific unit delivers both planned and on-call fires. (On-call fires are to the ship what targets of opportunity are to artillery units.) A fire control party with the supported unit conducts and adjusts on-call fires; they can be adjusted by an naval gun fire air spotter.
- General support. General support missions are assigned to ships supporting units of ABN IBCT size or larger. The fires of the general support ship are adjusted by an aerial observer or the liaison officer assign the fires of the ship to a battalion supporting arms liaison team for fire missions. Completing the mission, the ship reverts to direct support.

Coordination and Control Measures

5-67. Coordination and control measures that apply to naval gun fire are the same as for field artillery except for adding the terms fire support area and fire support station. They are as follows:

Chapter 5

- Fire support area. The fire support area is a sea area within which a ship can position or cruise while firing in support. It is labeled with the letters "FSA" followed by a Roman numeral–for example, FSA VII.
- Fire support station. The fire support station is a specified position at sea from which a ship must fire; it is restrictive positioning guidance. It is labeled with the letters "FSS" and followed by a Roman numeral—for example, FSS VII.

AIR OPERATIONS

5-68. Air operations normally are conducted using centralized control and decentralized execution and are integral to an airborne operation. They are performed concurrently and are mutually supporting and are planned with ample communications for liaison and control. The following paragraphs discuss counterair operations, close air support, air interdiction, airborne intelligence, surveillance, and reconnaissance, airlift missions, specialized tasks, and air operations planning in support of airborne operations.

COUNTERAIR OPERATIONS

5-69. The ultimate objective of counterair operations is to gain and maintain theater air supremacy. This has two purposes. It prevents enemy forces from effectively interfering with friendly areas and activities, and it precludes prohibitive interference with offensive air operations in the enemy area. This is accomplished by destroying or neutralizing the enemy's air offensive and defensive systems. (Refer to JP 3-01 for more information.)

Offensive Counterair Operations

5-70. Offensive counterair operations are conducted to seek out and neutralize or destroy enemy air forces at a chosen time and place. They are essential to gain air supremacy and to provide a favorable situation for other missions. Typical targets include—

- Enemy aircraft.
- Airfields.
- Tactical missile complexes.
- Command and control facilities.
- Petroleum, oil and lubricants and munitions storage facilities.
- Aircraft support equipment and their control systems.

Suppression of Enemy Air Defense

5-71. Suppression of enemy air defenses (SEAD) is conducted to neutralize, destroy, or temporarily degrade enemy air defense systems in a specific area by physical attack, electronic warfare, or both. Airborne electronic warfare assets (See ATP 3-36.) often are used in conjunction with other air operations or mission and are especially important to SEAD operations throughout the airborne operation. (Refer to ATP 3-01.4 for more information.)

Defensive Counterair Operations

5-72. Defensive counterair operations contribute to local air control by countering enemy offensive actions. By countering enemy offensive actions, theater forces can effectively use an in-place and operational radar warning and control system, consisting of both ground and airborne elements. They integrate and control the employment of fighters and air defense artillery.

5-73. Counterair tasks that are employed as a part of offensive and defensive counterair operations include air-to-surface attacks, fighter sweeps, and the protection warfighting function (escort). Field artillery may conduct counterair missions by striking enemy aircraft bases, helicopter forward arming and refuel points, and missile storage and launch sites.

CLOSE AIR SUPPORT

5-74. The objectives of close air support are to support surface operations by attacking hostile targets close to friendly surface forces. Each air mission requires detailed integration with those forces. (Refer to JP 3-09.3 for more information.)

AIR INTERDICTION

5-75. The objectives of air interdiction are to delay, disrupt, divert, or destroy an enemy's military potential before it can be brought to bear effectively against friendly ground forces. These combat operations are performed far enough away from friendly surface forces so that detailed integration of specific actions with the maneuver of friendly forces is not possible or required. (Refer to JP 3-03 for more information.)

AIRBORNE INTELLIGENCE, SURVEILLANCE, AND RECONNAISSANCE

5-76. Airborne intelligence, surveillance and reconnaissance missions are directed toward satisfying the requirements of joint force and component commanders engaged in surface and air operations within the joint operational area. The JFC's J-2 and J-3 jointly develop an overall collection strategy and posture for the execution of the intelligence, surveillance, and reconnaissance missions. These airborne missions provide timely information, either visually observed or sensor recorded, from which intelligence is derived for all forces. Surveillance operations continuously collect information; reconnaissance operations are directed toward localized or specific targets. (Refer to JP 3-30 for more information.)

AIRLIFT MISSIONS

5-77. The basic mission of airlift is passenger and cargo movement. This includes combat employment and sustainment, combat airlift missions that rapidly move forces, equipment and supplies from one area to another in response to changing conditions. Within airborne operations, combat employment missions allow a commander to insert surface forces directly and quickly into a objective area or airhead and to sustain combat operations through air landing of follow-on forces and sustainment operations. (Refer to JP 3-17 for more information.)

SPECIALIZED TASKS

5-78. Specialized tasks are those operation conducted in direct or indirect support of primary air and ground missions. These activities include, but are not limited to electronic combat, combat search and rescue, and air refueling operations.

AIR OPERATIONS PLANNING

5-79. Planning for air operations begins with understanding the JFC's and ABNTFC's mission and intent, and occurs in a collaborative manner with other components to integrate operations across the joint force. The joint air operations plan is the JFC's plan to integrate and coordinate joint air operations and encompasses air capabilities and forces supported by, and in support of, the airborne operation and other joint force components in the objective area.

Command and Control of Joint Air Operations

5-80. With the beginning of air operations in the objective area, provision must be made for command and control air operations with the supported ground effort. A joint operations center, where the supporting air component and the airborne force is represented, performs the planning, integration, direction, and supervision of the air effort according to the needs of the airborne force. Command and control considerations for air operations during airborne operations include:
- Preparation. If an airborne operation includes tactical air elements, the joint force commander directs part or all of the preliminary air efforts while other preparations for the operation are completed. Air support before and during the mounting of an airborne operation is a USAF responsibility. When the airborne force does not include tactical air elements. Therefore,

requests from the joint airborne force commander involving both reconnaissance and fire missions are processed through normal joint operations center channels.
- Assault. Requirements during the assault phase are the same for all airborne operations. During the dropping or air landing and assembly of assault elements, aircraft that are on air alert status over drop zones or landing zones defend against hostile surface or aerial reaction to the assault.
- Consolidation and exploitation. Air control network facilities in excess of tactical air control parties and airborne platforms are meager until the air landing of more supplies and reinforcements during this phase. In an operation, that does not involve an immediate linkup after seizure of objectives; the air landing of reinforcing or supporting elements provides for the rapid expansion and improvement of tactical air control networks to meet the needs of anticipated emergencies.
  - Aircraft providing support subsequent to the assault phase are based within the objective area, outside the objective area, or both. In view of the logistics demands of aircraft, air support is based within the objective area only when it cannot be effectively provided from outside. Limitations in the effective radius of aircraft are the determining factors. An airstrip or sufficiently adaptable terrain is one of these factors in selecting an objective area.
  - A single commander in the objective area has command over both ground and air elements. However, an officer charged with broader responsibilities whose headquarters is outside the objective area can retain such command.

Integration to Support Ground Operations

5-81. Adequate air support of an airborne operation requires some integration of airborne forces and air activity in support of ground operations.
- The air operations plan is based on the overall USAF mission and the amount of available strategic, tactical, and airlift effort. The effect of forecasted weather en route and in the proposed area of operation must be considered.
- Offensive and defensive air operations must be continuously planned in support of an objective area. Immediate air support must be continuously available (on air alert) in spite of an apparent absence of targets.

Air Traffic Control

5-82. Air traffic control in the airhead is initially an USAF combat control team responsibility. Augmented combat control teams can be replaced by Army air traffic control units at a later time. Air traffic services provided to airborne forces come from contingency corps assets. Liaison, beacon, and tower teams are the most frequently employed elements. Actions are as follows:
- During alert, marshalling, and deployment, a liaison team is sent to the headquarters that is planning the operation; it serves as a part of the section and provides advice on airspace management, especially in the airhead. The main concern in planning is the handoff between combat control teams and air traffic control parties. Combat control teams control the airhead with the advice and assistance from air traffic control personnel until follow-on air traffic control elements arrive.
- Beacon and tower teams deploy with the aviation or ABNAF main command post attached to the S-3 section. These teams provide initial air traffic control in the airhead. The beacon team provides terminal guidance for Army aircraft from their intermediate staging base into the airfield. The tower team augments the combat control team party. The amount of control given up by combat control teams to air traffic control teams depends on the size of the airflow.
- Operational control of air traffic control assets usually passes to the senior aviation unit commander once he is established in the airhead. Forward arming and refueling point and aviation assembly area operations include air traffic control elements and services, as specified by the senior aviation unit commander.

Landing Plan

SECTION IV – ASSEMBLY AND REORGANIZATION

5-83. The success or failure of the mission can depend on how fast the airborne force can regain its tactical integrity. The first goal of an airborne assault must be to deliver and assemble all available combat power as rapidly as possible. The more rapidly assault force personnel assemble and reorganize as squads, platoons, and companies, the sooner they can derig equipment and conduct the operations plan as a cohesive force. How efficiently and rapidly this happens is a direct result of detailed planning, cross loading on assault aircraft, and assembly and reorganization on the drop zone.

ASSEMBLY

5-84. Because the assembly is a key to success, it must be as deliberate and simple as possible. Assembly is more than accounting for personnel; the commander must ensure the force has regained its tactical integrity, and is organized and prepared to fight as a combined arms team. The slower the force assembles, the more it risks failure. An airborne force's assembly plan consists mainly of the following tasks:
- Establish and secure assemble areas and/or assault positions.
- Place all organic and attached weapon systems into action as quickly as possible.
- Reestablish mission command (such as radio networks or reporting to higher headquarters).
- Assemble the force and account for casualties and stragglers.

CROSS LOADING FOR RAPID ASSEMBLY

5-85. Cross loading of key personnel and equipment is an important factor in rapid assembly. Careful attention is given to cross loading and includes the following actions:
- Personnel. Separate key personnel in case an aircraft aborts or fails to reach the drop zone. This prevents the loss of more than one key officer or noncommissioned officer of a unit.
- Heavy equipment drop loads. Always plan for the possibility that one or more heavy equipment drop aircraft aborts before it gets to the drop zone or the equipment streams in and becomes unserviceable.
- Individual equipment and weapons. Planners should separate radios, mortars, antitank weapons, ammunition bundles, and other critical equipment or supplies as much as possible. No like items of combat-essential equipment from the same unit should be on the same aircraft if possible.
- Paratroopers jumping additional equipment containers or weapons systems such as antitank weapons should be jumped at any position in the stick to support cross loading and assembly plans. The commander must make a risk assessment when locating paratroopers in the stick carrying this equipment.
- Risks to both the paratrooper and mission accomplishment are present. If the paratrooper falls inside the aircraft, the remainder of the personnel may not be able to exit on that pass. In addition, this equipment increases the risks of the paratrooper being towed outside the aircraft.

ASSEMBLY METHODS

5-86. Based on the mission variables of METT-TC, the ground force commander may elect to use one of the following methods:
- Assembly on the assault objective. This method may be used when speed is essential, the objective is lightly defended, or the enemy can be suppressed.
- Assembly on the drop zone. This method may be used when: the drop zone does not use follow-on forces, speed is not essential, and dismounted avenues of approach from the drop zone to the objective are available.
- Assembly adjacent to the drop zone. This method may be employed when the drop zone is to be used by follow-on forces or if the drop zone is compromised during the airborne assault.

ASSEMBLY AIDS

5-87. To speed up assembly after landing, airborne forces use assembly aids to orient themselves on the ground and to locate their unit's assembly area. Assembly aids help identify personnel, equipment, and

Chapter 5

points or areas on the ground. Units can use visual, audible, electronic, natural, or individual aids; for reliability and ease of recognition, units combine these. Operators of assembly aids land as close as possible to their assembly area so they can mark the area. An USAF combat control team or LRSC team may place assembly aids if the situation permits. Partisans, special operations forces personnel, or high-performance aircraft can deliver assembly aids. Whenever possible, regardless of the method chosen to emplace the aids, commanders should provide backup personnel, backup aids, and backup delivery means.

Control Posts

5-88. An assembly control post is established by a small party equipped with assembly aids, which moves after landing to a predesignated location to help assemble assault forces. (See figures 5-3 and 5-4.) Each drop zone or landing zone has a control post in or near the unit assembly area to coordinate and regulate assembly. No standard organization exists for control posts; their composition varies with the size of the parent unit, the number and type of assembly aids, the terrain, and the assigned mission.

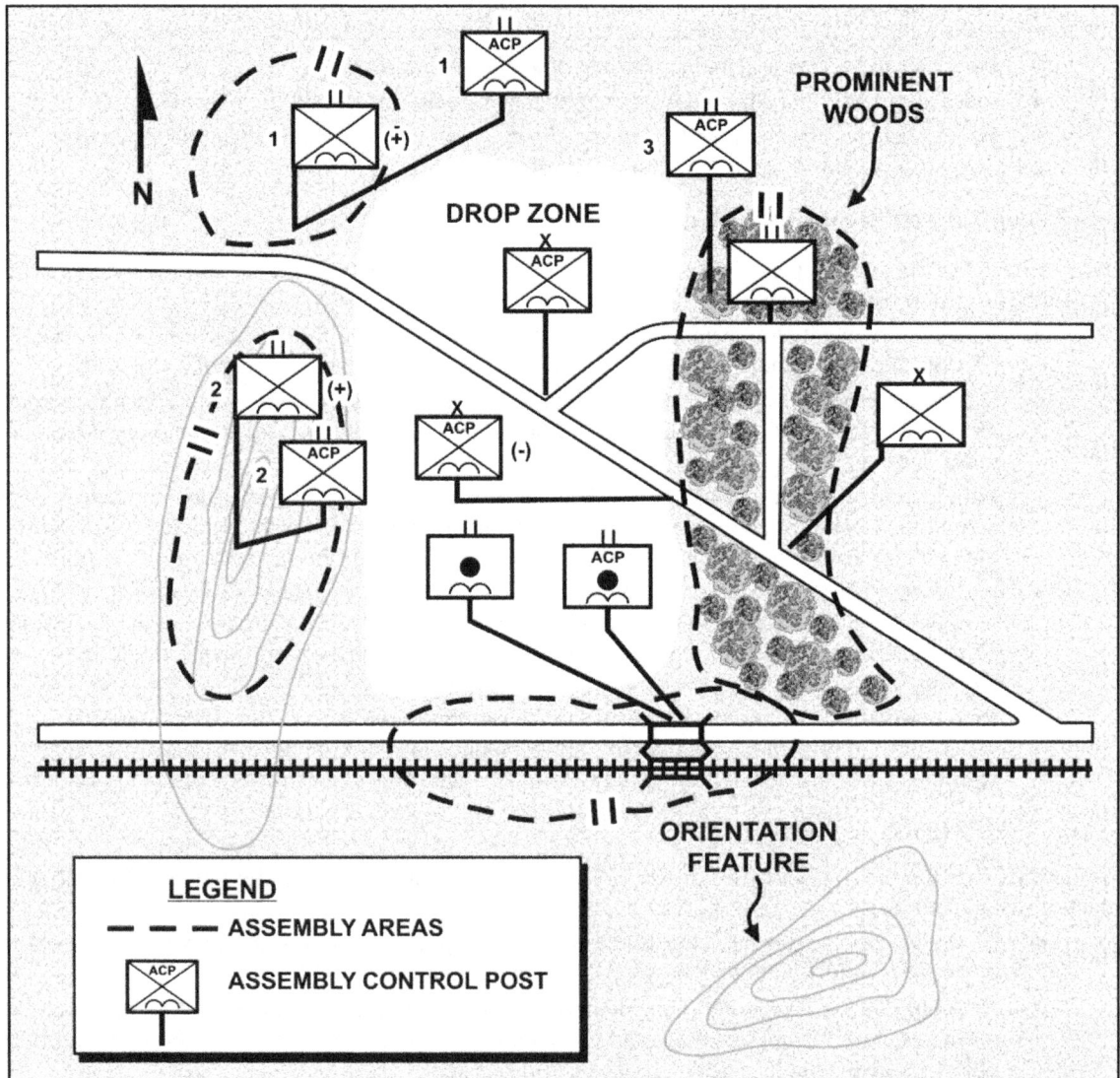

Figure 5-3. Assembly control posts for ABN IBCT forces landing on one drop zone

Landing Plan

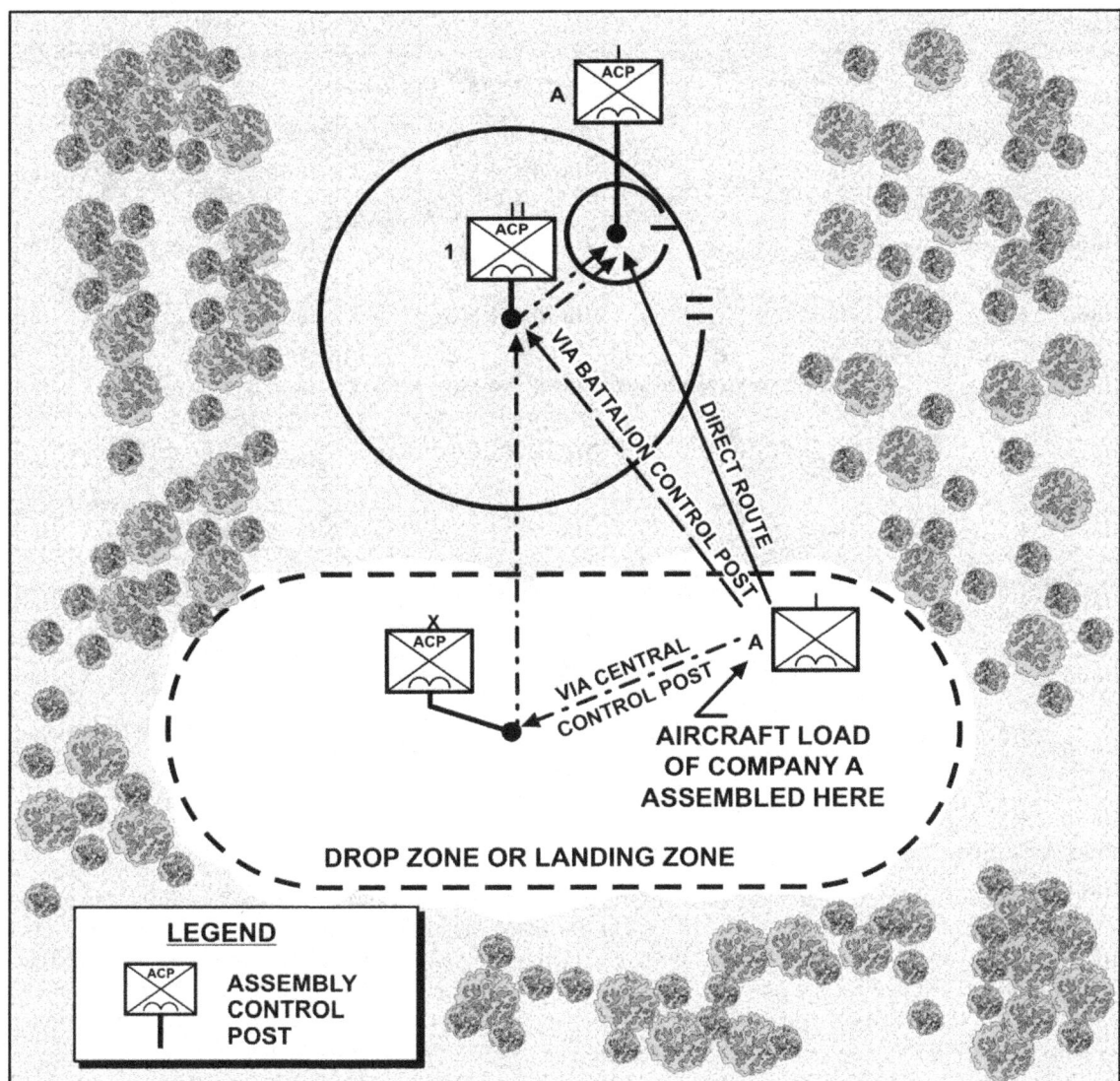

Figure 5-4. Movement of personnel to company assembly area

Line-of-Flight or Clock System

5-89. The line of flight parallels the parachute-landing pattern. This helps each paratrooper establish his own landing position relative to those of the other members of their planeload. Leaders use the clock system to brief personnel, calling the direction of flight 12 o'clock. (See figure 5-5, page 5-21.) After landing, personnel assemble to the right of the drop zone at 3 o'clock or to the left of the drop zone at 9 o'clock.

Chapter 5

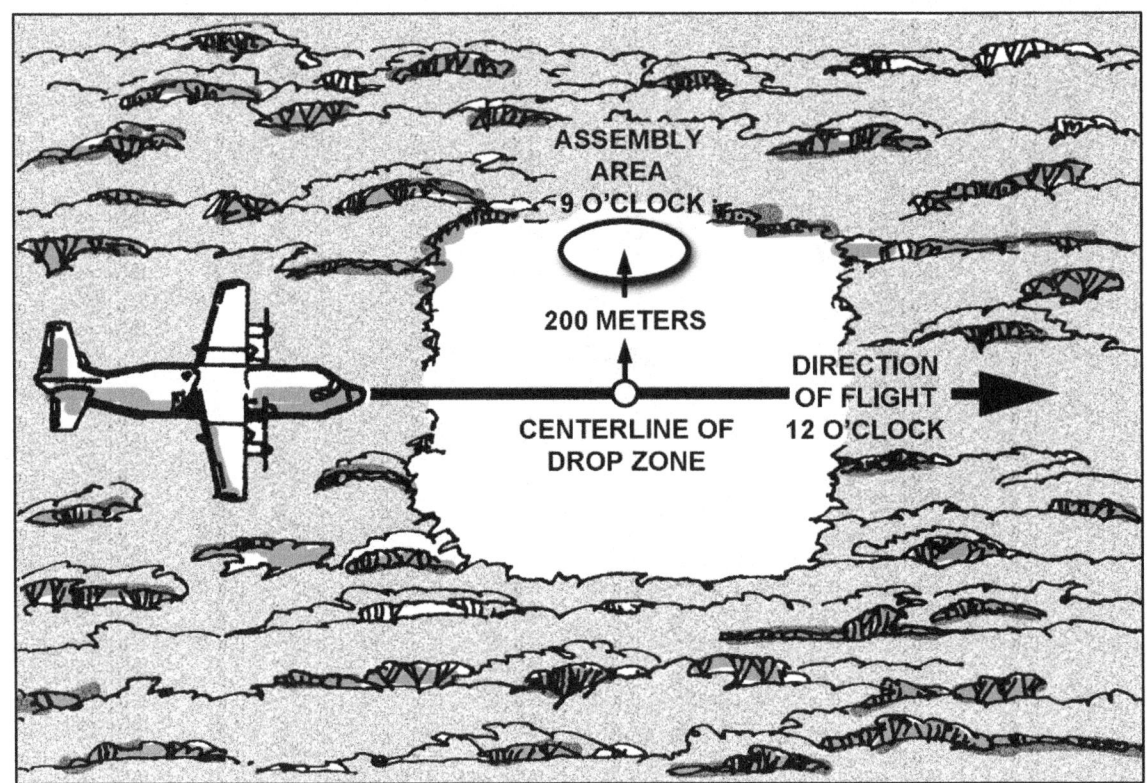

Figure 5-5. Line-of-flight/clock system

Natural Assembly Aids

5-90. These aids include landmarks or easily recognizable terrain features that forces can use as assembly aids or that personnel can orient their movement on. These features include hills; stream junctions; clumps of woods; or man-made objects like radio towers, bridges, buildings, crossroads, or railroads. Units cannot rely on natural features as the primary assembly aid. The assembly plan must be usable regardless of the drop zone. Executing contingency plans en route may require assembly on an alternate drop zone; an emergency exit from the aircraft can place paratroopers on an unfamiliar drop zone.

Assembly Equipment

5-91. Airborne forces carry visual, audible, or electronic aids to help them assemble. Planners assign different colors, sounds, and coded signals to each unit. The unit standard operating procedures standardizes assembly aids. However, units can adapt assembly aids to fit specific situations or environments. Terrain restrictions and battlefield noise do not restrict the use of the best assembly aids, which are simple to use. Units usually use visual assembly aids. Assembly equipment may include:

- Visual aids. Visual aids include visible light sources (such as beacons, flashlights, strobe lights, or signal mirrors; panels; flags), balloons; infrared lights (such as metascopes, flashlights with filters, infrared weapons sights, or starlight scopes); pyrotechnics; and chemical lights. These aids are simple to use and afford positive identification of assembly areas. However, the enemy can see them as well as friendly personnel can. The Stiner aid has a cloth panel with a colored letter; that is, HHC=White "H." They are the same color for night use. (See figure 5-6.) It has pockets for chemical lights; the letter and pockets are on both sides. It is mounted on a sectional aluminum pole that fits into a weapons case.

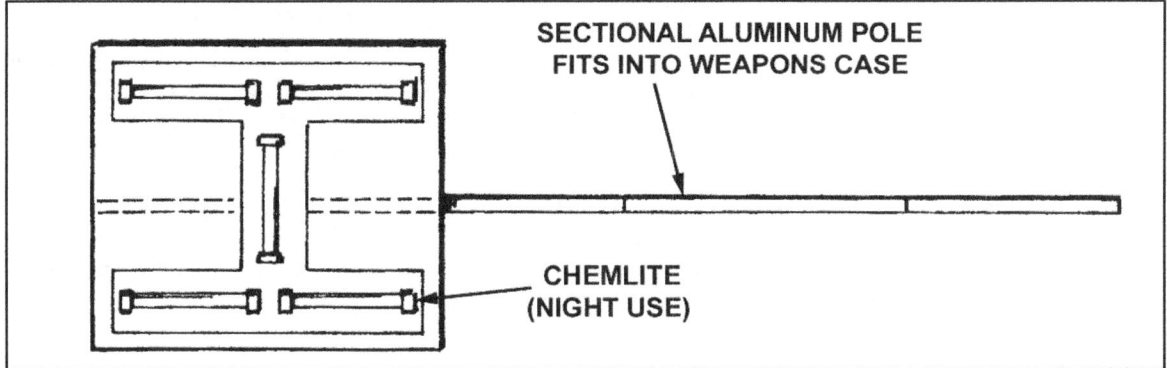

Figure 5-6. Stiner aid

- Audible aids. Audible aids especially help small units assemble at night. They include tin crickets, sirens, cowbells, air horns, triangles, dinner bells, ratchets, drums, gongs, whistles, bugles, and voice signals. They are used to identify individuals or assembly areas. Strong winds, gunfire, aircraft sounds, an elevation high above sea level, and other factors can limit their effectiveness. The normal sounds of the battlefield easily mask or confuse the sounds of audible assembly aids.
- Electronic aids. Units can use organic radios to direct small units to assembly areas, using landmarks as references. They can use radio homing devices. A homing device is a lightweight attachment to a standard field radio; it is an excellent aid for day or night assembly. With it, a radiotelephone operator (RATELO) can pick up a coded signal beam from a transmitter at the assembly area. By following the beam, the RATELO homes in on the transmitter and leads the unit to the assembly area. Signal crews can make equipment homing assembly aids from a standard portable field radio or transmitter. The unit attaches the radio to the equipment and turns it to a designated frequency. They encase it in shock-absorbing material just before its extraction from the aircraft. This method is especially useful for assembling crews on heavy equipment drop loads.
- Field-expedient aids. The unit uses numerous field-expedient assembly aids; only the Soldiers' imaginations limit the choices. For example, they can burn gasoline-soaked sand in cans or other containers; fashion a light gun or a one-direction light source by placing a flashlight in the receiver of a grenade launcher or other tube system; or lift a deployed main or reserve parachute so other unit members can see it.

Identification Markings

5-92. Identification markings aid positive, rapid identification of personnel and equipment that speed up a unit's assembly. The airborne force standardizes personnel and equipment markings for all subordinate units. Specifics on markings are as follows:
- Personnel markings. Soldiers use individual assembly aids to help recognize individuals and assemble units. Individual aids include colored armbands or helmet bands, distinctive patches or designs on uniforms, or helmet bands of luminous tape. Sortie commanders and key leaders, jumpmasters, safeties, other jumpmaster personnel, medics, and bump personnel use distinctive markings. Larger force standard operating procedures prescribe unit designations to prevent duplication and to allow unit-wide understanding.
- Equipment markings. Aerial equipment containers are identified by simple, distinctive markings. Distinctive unit markings are prescribed by larger force (ABN IBCT or division) standard operating procedures to prevent duplication and to assist in recognition by other units. Unit codes are placed on the bottom and all sides of each container; they should be visible for at least 50 meters. Various color parachute canopies, container colors, luminous tape and paints, smoke grenades, homing devices and lights can be used with the containers to facilitate identification on the ground and in the air. Lights and homing devices attached to equipment can be activated

manually aboard the aircraft just before extraction, either by improvised timer, or on the ground by the first individual to reach the equipment.
- Assault aircraft markings. A simple code symbol (using various designs, colors, and combinations of letters and numbers) can be painted on both sides of the fuselage of assault force aircraft to identify the contents. This symbol is large enough to be seen, and it indicates not only the type of equipment contained in the aircraft but the unit to which the equipment belongs.

FORCE ASSEMBLY

5-93. Commanders base the assembly of the airborne force on a simple, flexible plan that adapts to any likely situation. Assault forces assemble speedily, silently, and without confusion.

Drop Zone Assembly

5-94. When moving to an assembly area first as opposed to an assault objective, to speed assembly on the drop zone, forces should—
- Establish an assembly control point. Location of this point is dependent upon unit scheme of maneuver. A designated assembly control point officer in charge or non-commissioned officer in charge accounts for all paratroopers as they report to the control point. He then confirms azimuth and distance to the assembly area, or if the situation changes gives a new azimuth and distance to an alternate assembly area.
- Establish the assembly area. Locate the force assembly area in a covered and concealed position off the drop zone. The first group to arrive in the unit assembly area should first establish local security, and then establish the predetermined assembly aid as soon as possible to assist in directing the rest of the unit to the assembly area.
- Depart the assembly area. Move out rapidly on assigned mission once enough combat power is established. Leave a small element in the assembly area to assemble and account for personnel that have not arrived for later linkup.

Troop Briefings

5-95. Assault force personnel must understand individual actions during assembly. Personnel briefings include the following information:
- Brief assault force personnel and rebrief them on the assembly plan for their unit and on those of other units scheduled to share the same drop zone.
- Use visual aids such as maps, aerial photos, terrain models, and sand tables.
- Warn personnel to resist guiding on what appears to be a prominent terrain feature on a map. Once they are on the ground, the terrain feature probably cannot be seen. This is especially true if personnel land on the wrong drop zone or on the wrong part of a drop zone.
- Use the clock directional system. Instruct personnel to orient themselves and the general location of the assembly area by the direction of flight.

Note. Regardless of the actual azimuth, the direction of flight is always 12 o'clock.

FACTORS AFFECTING ASSEMBLY

5-96. When preparing the assembly plan consider dispersion (such as the speed, altitude, and flight formation of airlift aircraft and their effect (along with wind) of personnel and equipment in landing. The resultant landing pattern significantly affects assembly, as does the drop zone length and width, the training level of the airborne forces and pilots, the enemy situation, and cross loading.

Dispersion

5-97. The extent of dispersion is the result of the airlift formation; type, speed, and altitude of the aircraft; number of serials; sequence of delivery weather conditions; and aircrew proficiency. Dispersion covers the following:

- The speed at which airplanes carrying paratroopers cross the drop zone affects the length of the landing pattern. The greater the speed, the greater the distance that is covered between the exit of each paratrooper, thus increasing the length of the landing pattern. Planes cross the drop zone as slowly as is safely possible, and paratroopers exit rapidly to reduce dispersion.
- As paratroopers descend, they drift with the prevailing wind, but usually not at a uniform rate for each paratrooper. They can pass through strata of varying wind direction and velocity that causes some dispersion within the force. Due to the wind and higher altitude, there is a great possibility of dispersion. Therefore, the aircraft cross the drop zone at minimum altitudes that are consistent with the safety of aircraft and paratroopers.
- In parachute landings, the width of the landing pattern of paratroopers and equipment is the approximate width of the aircraft formation at the time of the drop. Therefore, keep the formation tight as possible to keep paratrooper and equipment together. If possible, place a company or battalion in the flight formation so that all planeloads of the force land in a small pattern as close as practicable to the assembly area.

Drop Zone Visibility

5-98. Darkness, fog, haze, rain, brush, trees, and terrain affect drop zone visibility on the ground, and hence impact on assembly. Darkness complicates assembly— poor visibility and difficulty in identifying or recognizing assembly areas, control posts, personnel, and equipment. Although it provide concealment, darkness contributes to confusion, to stragglers, and to the loss of equipment. An assembly during darkness takes longer and requires more elaborate assembly aids and larger control posts than a daylight assembly.

State of Proficiency

5-99. One of the most important factors that affect rapid assembly is proficiency. Assault forces must conduct parachute assaults and assemble as they would in combat. For specific missions, previous training is built on and tailored through detailed briefings to include maps, photos, and terrain models. When possible, rehearsals using assembly methods planned for the assault are used. Thorough orientation, rigorous training, aggressive leadership, and individual initiative have the single greatest impact on an airborne assault and assembly.

Enemy

5-100. Enemy action can have both a direct and indirect effect on assembly. Enemy action indirectly affects en route airlift capability to deliver the force to the correct drop zone. It directly affects friendly assembly once on the ground. Enemy opposition during or immediately after landing is a critical consideration affecting assembly due to the unusual vulnerability of the airborne force between landing and the completion of assembly or seizure of assault objectives. While the commander should attempt to achieve an unopposed landing, every possible provision is made to deal with enemy resistance. This requires accurate intelligence, responsive fires, and effective OPSEC and military deception. When assembly is on the assault objective, individuals linkup as they move using appropriate movement techniques in accordance with the operation plan.

MULTIPLE POINTS OF IMPACT

5-101. A slightly more complicated but more efficient method to facilitate rapid assembly of paratroopers, container delivery systems, and heavy equipment drop loads is the use of multiple points of impact. The theory and methods of cross loading apply as much to this method of delivery as to any other. When the USAF drops paratroopers along a single track (line of flight) down the center of the drop zone they use just one personnel and one heavy equipment drop impact point. When conducting multiple points of impact, USAF aircraft fly multiple tracks across the drop zone and use multiple impact points on the drop zone. By efficient cross loading, selecting assembly areas, and carefully selecting personnel and equipment points of impact, personnel, units, and equipment can be delivered closer to assembly areas than the single-track, one personnel and one heavy equipment drop point of impact method. (Refer to FM 3-21.38 for more information.)

Chapter 5

KEYS TO RAPID ASSEMBLY

5-102. Rapid assembly results from well thought out and rehearsed cross loading to include heavy equipment drop loads, and a thorough, but simple plan that applies for all drop zones. Take the following actions:

- Use the clock system (not magnetic azimuth system nor north, south, east, west) for direction or route to the assembly area.
- Use assembly areas that are easy to find without complicated assembly aids, even if dropped on the wrong part of the drop zone or on an unplanned drop zone.
- Be prepared with both day and night assembly aids, especially for drops scheduled at dawn or dusk.
- Locate assembly areas as close as possible to where personnel land.
- To permit rapid assembly, never locate assembly areas at either end of the drop zone. (Paratroopers should not have to walk from one end to the other.)
- Use personnel, unit, and equipment markings to speed assembly.
- Brief to all units what marking the same serial will use.
- Use sand tables extensively to brief on and rehearse assembly procedures as often as possible.

ASSEMBLY AREA ACTIVITIES

5-103. Not only do forces assemble as quickly as they can, but they get out of the assembly area as quickly as possible. They remain in the assembly area only long enough to establish mission command, task organize for further operations, and to determine the status of assembly. They modify plans as needed to meet changes in the situation and issue orders as appropriate.

5-104. Subordinate units determine minimum forces necessary to depart assembly areas early in the planning process. As the unit assembles on the drop zone, it immediately moves to its assault objectives once minimum force is established.

Departure From the Assembly Area

5-105. Battalion assault forces proceed on their assigned mission when assembly is complete or on order of the ABN IBCT commander. Reorganization of an assault force is complete when all subordinate units are assembled and command and fire control communications channels are established. As a result of inaccurate landings, enemy action, or assembly delays, assault forces may have to attack before assembly is complete. The ABN IBCT commander or acting commander makes this decision. In the absence of other orders, the battalion commander decides when enough of his battalion assault force has assembled to accomplish the mission. The time or conditions for assault forces to move out on their assigned missions are ordinarily established in the operation plan by higher headquarters.

Reports

5-106. Because of the dispersion of personnel and equipment in landing, the possibility of inaccurate landings, and the potential loss of aircraft during the air movement, commanders at all levels must learn the status of personnel and equipment in their units as soon as possible after landing. They need this information to determine combat power before executing the ground tactical plan. Reporting actions may include:

- All forces report their personnel and equipment status to the next higher unit at predetermined times or intervals until reorganization is complete. These status reports indicate the location of the unit; the number of personnel assembled and the number of known casualties; the number and type of crew-served weapons, vehicles, radios, and other recovered key equipment; and information available on missing personnel and equipment. Forces make abbreviated status reports from the drop zone as soon as they establish radio communications in accordance with standard operating procedures.

Landing Plan

- As personnel arrive in assembly areas, units make status reports by squad, platoon, and company. As commanders establish command posts in the assembly areas, they receive status reports from within the battalion by radio, messenger, or direct contact between commanders.

Security

5-107. Assault forces of the airborne operation are responsible for their own security once on the ground as they are vulnerable to enemy attack from all directions during assembly. For this reason, and because of the size of drop zones or landing zones, protection requirements are great in comparison with the size of the airborne force. Assault force actions may include:

- In small-unit drops, jumpmasters or chalk leaders may provide local security as their plane loads assemble and recover equipment dropped in aerial delivery containers.
- Leaders of air-landed personnel provide local security while the equipment is off loaded.
- Personnel linkup as the move to assembly areas using movement techniques as the situation dictates for security as well as speed and control.

REORGANIZATION

5-108. *Reorganization* includes all measures taken by the commander to maintain unit combat effectiveness or return it to a specified level of combat capability (FM 3-90-1). Reorganization is addressed in the landing plan and starts during assembly in assembly areas. If the enemy situation permits, assault forces may assemble on assault objectives after landing. Planning considerations for reorganization include—

- Designation and location of unit assembly areas and/or assault objectives.
- Use of assembly control posts and assembly aids.
- Establishment of command and fires communications networks.
- Reporting requirements.
- Limited visibility.
- Security operations.
- Reconnaissance and surveillance tasks.
- Coordination and final preparations before the seizure of assault objectives.
- Minimum force requirements to conduct mission.
- Time or conditions for assault forces to move out on missions.
- Designated personnel remain on the drop zone or landing zone to:
  - Assemble and provide security for stragglers.
  - Care for casualties.
  - Complete the recovery of accompanying or delayed equipment and supplies.
  - Establish straggler control and later linkup with the main force.

This page intentionally left blank.

Chapter 6
Air Movement Plan

After development, backbriefs, and approval of the landing plan, planners begin to finalize the air movement plan. This plan is the third step in planning an airborne operation and supports both the landing plan and the ground tactical plan. It provides the required information to move the airborne force from the departure airfield to the objective area. The plan includes the period from when forces load until they exit the aircraft. The air movement plan is a tab to the airborne operations appendix within Annex C, Operations.

SECTION I – JOINT PLANNING

6-1. Airborne assaults are inherently joint operations with specific service component responsibilities. Delivery by USAF airlift will be from either the continental United States, an intermediate staging base, or a theater airbase. Although the commander, airlift force is solely responsible for executing the air movement phase (See JP 3-17.), the ABNTFC normally exercises responsibility for the airlift plan, to include priority of airdrop and airland sorties, the preparatory fires plan, and the ground tactical plan in the airhead.

6-2. The ABNAF contributes its landing plan and the procedures for controlling and positioning personnel at the departure airfield. The airlift force controls takeoff times and, based on the ABNAF's landing plan, coordinates air operations and timing between different departure airfields, to ensure the proper arrival sequence at the drop zone, and/or landing zone (See JP 3-30.). The airlift force designates rendezvous points and develops the flight route diagrams. The ABNTF operation plan and ABNAF ground tactical plan determine flight routes and orientation of drop zones and landing zones.

SECTION II – ELEMENTS OF THE AIR MOVEMENT PLAN

6-3. The movement plan includes the period from when forces load until they exit the aircraft. Elements of the air movement plan include—
- Air movement table.
- Types of movement.
- Aircraft requirements.

AIR MOVEMENT TABLE

6-4. The air movement table forms the principal part of the air movement plan with the following information:
- Departure airfield for each serial.
- Number of aircraft for each serial.
- Chalk numbers for each aircraft, each serial, and each departure airfield; aircraft tail numbers correspond to aircraft chalk numbers.
- Unit identity of the airlift element.
- Name/rank of each USAF serial commander.
- Number and type aircraft.
- Employment method for each aircraft.
- Army force identity.

Chapter 6

- Name and rank of each Army commander.
- Load times.
- Station times.
- Takeoff times.
- Designated primary and alternate drop zones for each serial.
- P-hour for the lead aircraft of each serial. (Given in real time.)
- Remarks such as special instructions, key equipment, and location of key members of the chain of command.

6-5. Besides the air movement tables, the air movement plan contains the following information—
- Flight route diagram.
- Serial formation.
- Air traffic control.
- Concentration for movement.
- Allowable cabin/cargo loads.
- Airfield/forward logistic site aircraft maneuver on ground space.
- Aircraft parking diagram.
- Army personnel and equipment rigging areas at the departure airfield.
- Army control procedures during preparation for loading.
- Emergency procedures to include survival, evasion, resistance and escape and search and rescue planning.
- Weather considerations.
- Joint suppression of enemy air defenses, counterair, and air interdiction considerations.

TYPES OF MOVEMENT

6-6. The type of movement administrative or tactical must be considered when determining how to load the aircraft. Airborne forces can conduct an administrative movement to an intermediate staging base or remote marshalling base, and then transload into assault aircraft by using tactical loading. They are as follows:
- Administrative movements are non-tactical. Personnel and equipment are arranged to expedite their movement and to conserve time and energy. Economical use is made of aircraft cabin space, and planners make maximum use of the allowable cabin load (ACL).
- Tactical movements are when personnel and equipment are organized, loaded, and transported to accomplish the ground tactical plan. The proper use of aircraft ACL is important, but it does not override the commander's sequence of employment.

AIRCRAFT REQUIREMENTS

6-7. When the airborne force deploys, planning guidance from higher headquarters indicates the type of aircraft available for the movement. Based on this information, the unit commander determines and requests the number of lifts by the type of aircraft required to complete the move. The air movement planner must ensure that each aircraft is used to its maximum capability. This is based on the information developed on unit requirements, ACLs, and available passenger seats. The methods of determining aircraft requirements are the weight method and the type-load method. They are as follows:
- Weight method. This method is based on the assumption that total weight, not volume, is the determining factor. Since aircraft sometimes run out of space before exceeding the ACL, this method is no longer widely used. It has been replaced by the type-load method. The long distances involved in reaching an objective area, the necessity of the aircraft to circle for extended periods before landing, and the large amounts of fuel needed to sustain the aircraft can result in the aircraft having to reduce its ACL. As a rule, the longer the deployment, the lower the ACL.

- Type-load method. In a force air movement, a number of the aircraft loads contain the same items of equipment and numbers of personnel. Identical type loads simplify the planning process and make the tasks of manifesting and rehearsing much easier. Used for calculating individual aircraft lift requirements, the type-load method is the most common and widely accepted method of unit air movement planning. It requires consideration of load configuration and condition on arrival at a desired destination, rapid off-loading, aircraft limitations, security requirements en route, and the anticipated operational requirements. Use the more detailed type-load method in planning force movements.

SECTION III – LOAD PLANNING CONSIDERATIONS

6-8. When preparing the air movement plan, the S-3 Air considers tactical integrity, cross loading, self-sufficiency of each load, and chance of executing the air land option. The S-3 air keeps units intact as much as possible. For airborne operations, this can mean placing forces larger than squads on separate aircraft so they exit their respective aircraft over the same portion of the drop zone. This facilitates rapid assembly by placing units close to their assembly areas on landing.

TACTICAL INTEGRITY

6-9. Maintaining tactical integrity includes the following guidance:
- The S-3 Air must understand and develop the tactical cross load based on the ground tactical plan.
- Key leader or staff groups are cross loaded and positioned within the assault echelon to best control ground maneuver and provide redundancy of mission command for the airborne operation.
- Fire support teams and their radiotelephone operators should be on the same aircraft with the commander they support; they should jump to land next to him.
- Platoon leaders (and platoon sergeants on different aircraft) should have their forward observers, radiotelephone operators, at least one machine gun crew, and one Javelin gunner on the same aircraft.
- Each aircraft has at least one unit non-commissioned officer or commissioned officer for each unit with personnel on board. Each aircraft has Army leadership present.
- To ensure tactical integrity, distribute the company commander, unit first sergeant, and executive officer in different aircrafts.

CROSS LOADING

6-10. Cross loading distributes leaders, key weapons, and key equipment between the aircraft of the formation to avoid total loss of mission command or force effectiveness if an aircraft is lost. Give careful attention to cross loading during rapid assembly.

6-11. Separate key personnel in case an aircraft aborts or fails to reach the drop zone. This prevents the loss of more than one key officer or NCO of a unit. Properly planned cross loading accomplishes the following:
- Personnel from the same unit land together in the same part of the drop zone for faster assembly.
- Equipment or vehicle operators and weapon system crews land in the same part of the drop zone as their heavy-drop equipment so they can get to it, derig it, and put it into operation quickly.
- If one or more aircraft abort either on the ground or en route to the drop zone, some key leaders and equipment still are delivered.

6-12. When planning airborne force cross loading, remember—the fewer key people on the same aircraft, the better. If possible, separate the following personnel:
- The ABN IBCT commander from his executive officer and battalion commanders.
- The battalion commander and his company commanders.
- The brigade executive officer, and the brigade S-3.
- The battalion executive officer, and S-3 from the same battalion.
- The primary ABN IBCT and battalion staff officers and their assistants.
- The company commander, executive officer, and first sergeant from the same company.

Chapter 6

- The platoon leader and platoon sergeant from the same platoon.

6-13. Always plan for the possibility that one or more heavy equipment drop aircraft aborts before it gets to the drop zone or the equipment streams in and becomes unserviceable. Take the following actions:
- Cross load heavy-drop equipment to have the least possible impact on the mission if it does not arrive in the drop zone. Separate critical loads so if an aircraft aborts or fails to reach the drop zone, no single unit loses more than one key officer or non-commissioned officer or a significant proportion of the same type of combat-essential equipment.
- Coordinate closely with the Air Force so heavy-drop equipment loads are loaded in the reverse order they should land.
- Do not include the same type of critical equipment from the same unit, or like equipment from different units in the same aircraft loads. This applies whether it is to be airdropped or air-landed.
- Cross load heavy-drop equipment in one of the following ways:
  - Select heavy equipment points of impact to support the ground tactical plan. Place loads so they land close to the location where they are used.
  - Cross load the parachutists to first support the ground tactical plan; then coordinate their landings with those of the heavy-drop platforms.
  - Coordinate the selected heavy equipment point of impact for each load with the Air Force mission commander, when using multiple heavy equipment points of impact.
  - Do not load two or more like platforms from the same unit on the same aircraft because the aircraft are moving too fast to drop more than one platform in the same area.
- Separate radios, mortars, antitank weapons, ammunition bundles, and other critical equipment or supplies as much as possible. No like items of combat-essential equipment from the same unit should be on the same aircraft. Apply the following:
  - A weapons system should be loaded on the same aircraft as its crew.

Note. Only one crew-served weapons squad/team should be on each aircraft.

  - A RATELO should jump the same aircraft as the leader he supports, either just before or after him. Another good method is for the leader to jump the radio himself. In this way, he still can set up immediate communications even if he and his RATELO separate on the drop zone.
  - The container, weapon, and individual equipment can and should be jumped at any position in the stick to support cross loading and assembly plans. The commander makes a risk assessment when locating paratroopers in the stick carrying this equipment. Risks to both the paratrooper and mission accomplishment are present. If the parachutist falls inside the aircraft, the remainder of the personnel may not be able to exit on that pass. This equipment increases the risks of the paratrooper being towed outside the aircraft.
  - Individual crew-served weapons (such as machine guns, mortars, antitank weapons) and other critical equipment or supplies should be distributed on all aircraft.
  - Communications equipment, ammunition, and other supply bundles must be cross loaded.

SELF-SUFFICIENCY

6-14. Each aircraft load should be self-sufficient so its personnel can operate effectively by themselves if other aircraft misses the drop zone, makes an emergency landing somewhere else, or aborts the mission. Take the following action:
- A single (complete) weapons system should have the complete crew for that system on the same aircraft along with enough ammunition to place the weapon into operation.
- For airland or heavy equipment drop operations, trailers and weapons are manifested with their prime movers.
- Squads should stay together on the same aircraft; fire teams are never split.

- Squads and fire teams should jump both aircraft doors to reduce the amount of separation on the drop zone.

SECTION IV – LOADING AND DELIVERY OF FORCES

6-15. The air movement plan contains the information required to ensure the efficient loading and delivery of forces to the objective area in the proper sequence, time, and place to support the ground tactical plan. The following paragraphs will discuss load planning sequencing, vehicle load planning, air movement planning worksheet, and aircraft utilization.

LOAD PLANNING SEQUENCE

6-16. Planners can best accomplish the movement of forces by air for an airborne assault by developing plans in an orderly sequence, such as—
- Preparing vehicle load cards.
- Preparing air movement planning worksheets for each unit (company through battalion).
- Preparing basic planning guides (company and battalion) and forwarding them to higher headquarters (battalion and ABN IBCT).
- Establishing priorities for entry into the objective area by echelon—assault, follow-on, and rear. Units establish priorities within each echelon to phase personnel and equipment into subsequent echelons if aircraft are not available.
- Preparing a force aircraft utilization plan to determine aircraft requirements and type loads.
- Preparing air-loading tables to facilitate rapid deployment.
- Complete and print draft load plans through the Integrated Computerized Deployment System (ICODES) to help identify issues prior to completing actual load plans.

6-17. Units receive their missions and review previous plans. Units—
- Amend the plans based on the task organization dictated by the ground tactical plan.
- Allocate available aircraft. If aircraft are not available, they phase low-priority items to the follow-on or rear echelon.
- Prepare air-loading tables and manifests.
- Prepare the air movement table.
- Prepare a DD Form 1387-2 (*Special Handling Data/Certification*) for hazardous materials.

LOAD PLANNING OF VEHICLES

6-18. Base vehicle load plans on standard operating procedures and mission tailoring. Then, update them according to aircraft availability and type.
- Heavy-drop vehicles are first loaded with as much unit equipment as they can hold. The vehicle's load capacity should not be exceeded, and all cargo must be secured in the vehicle's cargo compartment.
- Vehicles are measured and weighed after they have been loaded. Some items, especially ammunition, cannot be rigged on the vehicle, but can be carried as ballast on the platform.
- Vehicle load cards are made for each vehicle to be loaded aboard an aircraft. Each sketch includes such information as load data for the vehicle; length and width of the vehicle when the vehicle carries cargo; the names and locations of the cargo in the vehicle.

AIR MOVEMENT PLANNING WORKSHEET

6-19. The air movement planning worksheet is a consolidated list of a unit's equipment and personnel. It is not a formal DA form; it is an example of a locally made form. If necessary, use grid-type paper in lieu of a printed form. The worksheet lists all the dimensions and cargo loads of vehicles. It must include all on-hand equipment and personnel, and the full amount authorized by the unit table of equipment. Short items still are included as equipment, and personnel shortages can be filled if alerted for deployment. This prevents the need for constant revision of the worksheet. Basic loads of ammunition carried with the unit, which must be palletized or placed in door bundles, should be included.

6-20. The basic planning guide form is a report prepared by ground forces to determine the aircraft required for an airborne operation. The S-3 Air for the battalion collects the basic planning guide forms from the subordinate companies and consolidates them at battalion level. He submits them to U.S. Army riggers, airlift control element (ALCE), and the departure airfield control group (DACG), depending on the type of movement required.

UNIT AIRCRAFT UTILIZATION FORM

6-21. The unit aircraft utilization plan identifies equipment by aircraft load; this simplifies planning of identical types of loads. The goal is to support the ground tactical plan while maximizing efficient use of USAF assets. The first step is to weigh personnel and equipment by echelon. Then, add up the aircraft loads to determine how many aircraft are needed. If too few aircraft are available to meet the planned echelonment, this becomes readily apparent. At this point, priorities are applied and equipment and personnel are phased back to fit airlift constraints.

SECTION V – AIRCRAFT LOAD AND AIR MOVEMENT TABLE

6-22. The development of aircraft loads is accomplished through reverse planning. The planner must have a mosaic or facsimile when developing the heavy equipment point of impact, personnel point of impact, and personnel manifests. Aircraft loads must support the assembly and ground tactical plans through effective cross loading. This includes—

- Preparing the load. Using the mosaic, facsimile, or sketch, preparers mark the desired single or multiple heavy equipment points of impact for all equipment, and the personnel point of impact. Line off the sketch in 70-meter (75-yard) increments from the personnel point of impact. This represents the normal one-second parachutist interval. Make the set lines perpendicular to the line of flight so that the name of the parachutist associated with a particular piece of equipment can be marked on the sketch.
- Planning purposes. Heavy-drop equipment lands 400 yards apart on C-130 and C-17. The name of the parachutist who must obtain his equipment is entered on the line nearest the equipment. Personnel not associated with a particular piece of equipment can be marked on the lines closest to their assembly area. Take the personnel manifest directly off the drop zone schematic, the result is a manifest order that facilitates quick assembly.
- Allocating seats. Once the commander has developed the cross-load plan, he notifies involved units how many and which seats they have on each aircraft. Platoons can be manifested in multiple aircraft to facilitate cross loading, but personnel are placed in stick order on each aircraft to exit and land in the same general area on the drop zone.
- Making internal adjustments. Each company commander in turn cross loads his part of the split platoon within his part of the stick to best support the assembly plan and ground tactical plan. (See figure 6-1.)
- Preparing the manifest. Manifesting is accomplished in the reverse order of exit. (Refer to paragraph 6-26 of this publication for more information.)

Air Movement Plan

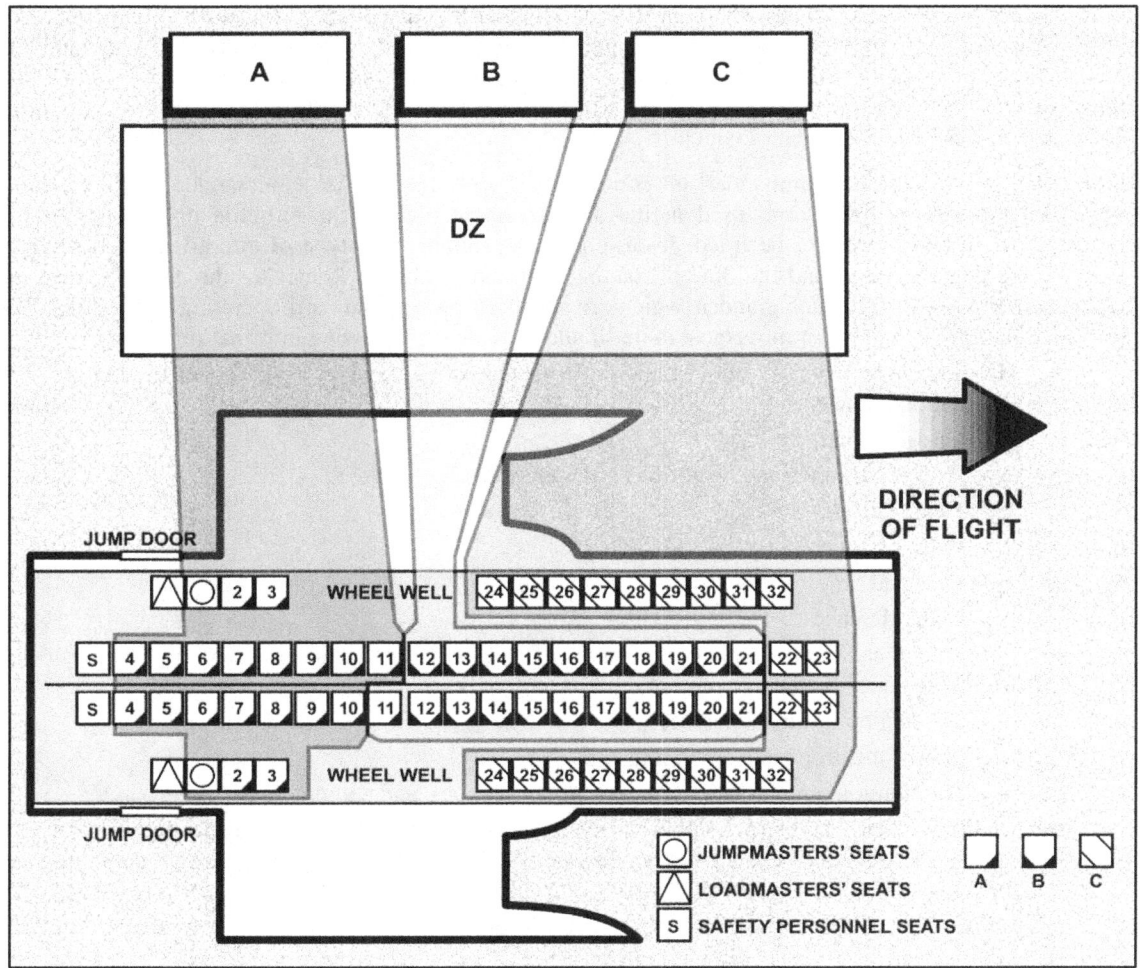

Figure 6-1. Cross-loaded aircraft

Chapter 6

6-23. The air movement table assigns units to serials within the air columns. Locating units in successive serials is according to priorities established for landing. Units maintain tactical integrity of Army and airlift units as far as practicable. All elements in a given serial land on the same drop zone or landing zone in the objective area; however, certain aircraft in a serial can continue on to drop reconnaissance and security forces in their planned areas of employment.

6-24. The ground forces commander in coordination with the USAF commander prepares the air movement table form. This form, used in the air movement plan to the airborne operations, allocates aircraft to the ground forces to be lifted. It designates the number and type of aircraft in each serial and specifies the departure area and the time of loading and takeoff. Exact format for the air movement table depends on the needs of the commander, which are specified by unit standard operating procedures. There is no specific format, but the air movement table should provide the following information:

- Heading. When the air movement table is published as a part of the order, the following elements are included:
  - Classification.
  - Appendix, annex and operations order number.
  - Headquarters.
  - Place of issue.
  - Date and time.
  - Map reference.
- Serial number. Serial numbers are arranged consecutively in the order of flight. Consider the following factors in the assignment of units to serials:
  - Mission of the airborne force.
  - Size of the drop zone or landing zone.
  - Distribution (cross loading) of personnel, weapons, and equipment.
- Chalk number. The chalk number specifies the position of aircraft being loaded in each serial. Loads are numbered sequentially according to serial numbering, such as Serial 1 contains Chalks 1 through 12; Serial 2 contains Chalks 13 through 24.
- USAF units. This section includes USAF information that is important to the ground force commander.
  - Airlift unit. This is the designation of the airlift unit that is transporting or furnishing the aircraft for each serial.
  - Serial commander. This is the senior USAF officer in the serial.
  - Number and type of aircraft. The exact number and type of aircraft that actually fly in the serial are shown in this column.
- U.S. Army forces. This section includes information directly related to the ground forces.
  - Aircraft required. The number of airplanes required to transport the force.
  - Employment. Type of movement (such as parachute, heavy equipment, container delivery system, or air-landed).
  - Unit loaded. The airborne force being loaded.
  - Serial commander. The senior airborne officer in the serial.
  - Departure airfield. Name or code name of the departure airfield.
  - Load time. The time established by the airlift and ground force commander to begin loading. Parachutists can require 30 to 45 minutes to load, depending on the aircraft and any accompanying equipment (door bundles, and parachutes for in-flight rigging). Heavy-drop and container delivery system loads should take about two hours for loading.
  - Station time. The time the passengers, equipment, and crew are loaded and ready for takeoff.
  - Takeoff. The time the aircraft is scheduled to depart from the airfield.
  - Aircraft formations. The type formation the aircraft will fly.
  - Objective. The name or designation of the drop zone, landing zone, or airfield.

- Time over target. Time over target is the time of arrival at the objective area.
- Direction of flight over the objective area.
* Other items. Other items that can be included in the air movement table (at the commander's discretion) are as follows:
  - Number of personnel by serial/chalk.
  - Initial and final manifest call times.
  - Prejump training times.
  - Type parachute.
  - Weather decision.
  - Weather delay.
  - Time for movement to the departure airfield.
  - USAF station time.
  - Remarks.

SECTION VI – MANIFESTS AND AIR-LOADING PLANNING SYSTEM

6-25. To help maintain order and sequence of delivery of personnel and equipment to the drop zone the ABNAFC and planning staff use a process of prioritization and accounting for the ABNAF personnel and equipment. Manifests and computerized systems stream line this process.

MANIFESTS

6-26. The flight manifest DD Form 2131 (*Passenger Manifest*) is an exact record of personnel by name, rank, Department of Defense (DOD) identification number, and duty position in each aircraft. It is a brief description of the equipment, with the station number, as loaded in the aircraft. Load computations for personnel and equipment are listed. Make a separate form for each aircraft.

6-27. Copies are made for the arrival and departure airfield control group, departure airfield control officer, pilot, and ALCE. The jumpmaster or senior Army representative on the aircraft retains a copy. The senior ground forces member or primary jumpmaster in each aircraft finalizes the form. The USAF authorizes it, and the ground force representative signs it after verifying the personnel on the manifest.

INTEGRATED COMPUTERIZED DEPLOYMENT SYSTEM

6-28. The Integrated Computerized Deployment System (ICODES) is a computer-based automated system designed to simplify the outload plans for combat forces.

6-29. Files data base. By computerizing the necessary loading characteristics, weight restrictions, and equipment configurations required to outload the airborne force, this system automatically tells the commander the load configurations and number of aircraft required to move a specific force. The files data base includes the following:

* USAF aircraft characteristics. This file includes all data that affect the placement of equipment on a particular aircraft; it contains data on the C-130, C-17, and C5A/B aircraft, which appropriate personnel update as changes occur.
* Items and uniform. This file contains size data on all the equipment in the unit that requires floor space. Commanders use the data for airland or airdrop. The file contains information about the aircraft center of balance, the psi of the tires, whether the item requires shoring or not, and whether the item can be turned or not. The unique feature of this file is that it considers inseparable items of equipment together (for example, a jeep and trailer or a high mobility multipurpose wheeled vehicle and 105-millimeter howitzer). This permits the program to load an item of equipment with its prime mover.
* Force package and options. This file contains 12 force packages and about 1000 modular force package options.

6-30. Commander input. The automated air load planning system allows commanders to input force packages, options, items, prime movers and the towed pieces of equipment, and multiples of each. Once a force package or option is entered into the program, the force package or option can be changed for the specific run of the program to meet mission requirements.

Chapter 7
Marshalling Plan

After the air movement plan is developed, backbriefed, refined, and approved, the next plan to be finalized is the marshalling plan. The previous three plans— ground tactical, landing, and air movement— are used to determine the number of personnel and vehicles to be stationed at or moved through each airfield. The marshalling plan provides the necessary information and procedures by which units of the airborne force complete final preparations for combat, move to departure airfields, and loads the aircraft. It provides detailed instructions for facilities and services needed during marshalling. This chapter assists airborne assault force (ABNAF), commanders and staffs in planning for marshalling and sustainment.

SECTION I – PREPARATION

7-1. *Marshalling* is the process by which units participating in an amphibious or airborne operation group together or assemble when feasible or move to temporary camps in the vicinity of embarkation points, complete preparations for combat, or prepare for loading (JP 3-17). The marshalling plan appears as a tab to the airborne operations appendix Annex C, Operations. S-4s (in coordination with S-3s), are the principals to the commanders for marshalling. Marshalling begins when elements of the force are literally sealed in marshalling areas and it terminates at loading. The marshalling plan is designed to facilitate a quick, orderly launching of an airborne assault under maximum security conditions in the minimum possible time.

7-2. Units complete the following preparations before marshalling as a minimum:
- N-hour sequence. As soon as a force is notified of an airborne operation, it begins the reverse planning necessary to have the first assault aircraft en route to the objective area at the predetermined time. The N-hour sequence contains preparation activities, along with planning actions that must take place within a flexible schedule, ensuring that the force is prepared and correctly equipped to conduct combat operations on arrival.
- Rehearsals. Rehearsals are conducted at every echelon of command. They identify potential weaknesses in execution and enhance understanding and synchronization. Full-scale rehearsals are the goal, but time constraints may limit them.
- Static load training (SLT). Requires coordination with USAF loadmasters. Scripted event that trains joint force for daytime and limited visibility aircraft egress and personnel and equipment on/off-load procedures.
- Sustained airborne training (SAT). Conducted no more than 24 hours prior to execution of the airborne assault. Paratroopers receive SAT from the jumpmaster and safety from the door from which they will exit. SAT normally is conducted during SLT. (See TC 3-21.220.)
- Assembly, inspection, and maintenance. As soon as feasible, units assemble the equipment and supplies that are to accompany them to the objective area. Hold inspections to determine the status of equipment. Perform maintenance and prepare parachutes, aerial delivery containers, and heavy equipment drop loads. Commanders and leaders brief personnel, and rations and ammunition are issued. Personnel eat as time permits.
- Storage of nonessential items. Individual clothing and equipment, and unit equipment not needed in the objective area are packed in suitable containers and stored with the rear echelon or rear detachment.

Chapter 7

> *Note.* At a minimum, marshalling activities include briefing personnel, inspecting, preparing airdrop containers, issuing rations and ammunition, and resting.

SECTION II – MOVEMENT

7-3. A marshalling area is a location in the vicinity of a reception terminal or pre-positioned equipment storage site where arriving unit personnel, equipment, materiel, and accompanying supplies are reassembled, returned to the control of the unit commander, and prepared for onward movement. (Refer to JP 3-35 for more information.) Unit marshalling areas should be located near departure airfields to limit movement, higher headquarters can either control the movement to the marshalling area completely, or it can get a copy of the march table and use it to control the traffic out of the assembly area, along the route of march, and into the marshalling area. Advance parties assign personnel to areas.

7-4. The S-4 of the unit to be marshaled notifies higher headquarters on the number of organic vehicles that the unit can give to move its personnel and equipment to the marshalling areas. This information and the personnel list furnished by the S-3 must be available early enough during planning to procure other transportation required for the movement.

7-5. When marshalling areas are on airfields, they are placed temporarily at the disposal of the airborne force's higher headquarters. The air base commander's permission is obtained by the tactical units that must conduct activities outside of the camp area.

7-6. Parachute issue and rigging may be conducted on the ramp, alongside the aircraft, or in-flight. Advantages and disadvantages are listed below in table 7-1.

Table 7-1. Parachute issue

Issue/Rig	Advantages	Disadvantages
RAMP	• Reduces the parachute supply problem. • Efficient use of personnel. • Supply accountability.	• Parachutists may require transportation to the aircraft. • Parachutists are rigged for a greater period.
PLANE SIDE	• Parachutists are not required to walk while rigged. • Decentralized execution reduces rigging time.	• Parachutes must be transported to the aircraft. • Rigging process may impede other activities.
IN-FLIGHT	• Prevents fatigue during long flights. • Provides more time for rehearsals and inspections.	• Reduces the number of parachutists that an aircraft can carry. • Requires loading of parachutes on the aircraft.

SECTION III – PROTECTION

7-7. Protection tasks and systems preserve the force so the commander can apply maximum combat power to accomplish the mission. Preserving the force includes protecting personnel and physical assets of the U.S., host-nation, and multinational military and civilian partners.

7-8. The marshalling area should be surrounded by security fencing or, at least, triple-strand concertina. It should have a posted security area outside the perimeter that is at least 50 meters wide and cleared of brush and trees. If available, use lights to illuminate the security area. Gates to the camp should be two lanes wide to accommodate heavy traffic.

PASSIVE DEFENSE MEASURES

7-9. Uncommitted airborne forces pose a strategic or operational threat to the enemy. Avoid concentrating forces during marshalling to keep impending operations secret and to deny lucrative targets to the enemy. Dispersal methods include the following:

- Move. Units move rapidly under cover of darkness to dispersed marshalling areas near air facilities.
- Control. Commanders control movement to loading sites so most personnel arrive after the equipment and supplies are loaded on the aircraft.
- Prepare. Commanders prepare for loading before arrival at the loading site.
- Avoid. Commanders avoid assembling more than 50 percent of an ABN IBCT at a single point.

DISPERSAL

7-10. The degree of dispersal is based on an intimate knowledge of the operation's problems and what is best for the overall operation. Regardless of the dispersed loading method, the airlift commander ensures that aircraft arrive over the objective area in the order required by the air movement plan. Depending on the situation, one of the following methods is used:
- Movement to departure air facilities. Move airborne forces and equipment to departure air facilities where airlift aircraft may be dispersed.
- Movement to intermediate staging base. Before the mission, airlift aircraft fly to an intermediate staging base to pick up airborne forces and equipment. Airlift airborne forces and equipment to dispersed departure airfields; the mission originates from these facilities.
- Combining methods. Airlift aircraft fly to intermediate staging bases for the equipment before the mission. The equipment is airlifted to the dispersed departure airfields and the mission originates from these facilities, or airlift aircraft stop en route at intermediate staging bases to pick up personnel. Crews load aircraft quickly, so the fewest possible aircraft are at the intermediate staging base at one time.

SECTION IV – DEPARTURE AIRFIELD-MARSHALLING AREA

7-11. Base the departure airfield selection on the proposed air movement and the capability of airfields to handle the traffic. Designate loading sites near departure airfields after selecting departure airfields.

SELECTING DEPARTURE AIRFIELDS

7-12. For a specific situation or operation, one or a combination of the following factors can determine the selection:
- Mission.
- Airfields (number required, location, and type).
- Runway length and weight-bearing capacity.
- Communications facilities.
- Navigational aids and airfield lighting.
- Locating participating units and marshalling areas.
- Radius of action required.
- Vulnerability to enemy action to include chemical, biological, radiological, and nuclear.
- Other air support available or required.
- Logistical support available, required, or both.
- Facilities for reception of personnel and cargo.
- Facilities for loading and unloading of personnel and cargo.
- Facilities for dispatch of personnel and cargo.
- Facilities to support rigging and storage of heavy-drop platforms.

Note. While dispersion is necessary to avoid vulnerability to enemy action, excessive dispersion increases control problems and can diminish the effectiveness of other supporting ground and air operations.

Chapter 7

SELECTING AND OPERATING MARSHALLING AREAS

7-13. The marshalling area is a sealed area with facilities for the final preparation of paratroopers for combat. Commanders select marshalling areas based on the air movement plan and other considerations. Another way to avoid concentration of personnel is to time-phase the movement of personnel from their home bases through the marshalling area to the departure airfield, minimizing the buildup of forces. After choosing the marshalling areas and departure airfields, choose loading sites near the airfields.

7-14. The following factors are considered when selecting marshalling areas:
- Distance to airfield(s).
- Time available.
- Current facilities.
- Availability of personnel and materials for construction.
- Availability or access of maneuver and training areas.
- Communications requirements.
- Briefing facilities.
- Locating participating units.
- Security or vulnerability to enemy action.
- Logistical support available or required.

7-15. In the marshalling plan, the S-4 (in coordination with the S-3), assigns units to marshalling areas near the departure airfields the units will use. Make every effort to locate the areas as close as possible to departure airfields to reduce movement time between them and to reduce requirement for vehicles.

7-16. The ABNAF's higher commander is responsible for the operation and maintenance of the marshalling areas. He includes the following:
- Provide operating detachments and necessary equipment for each area. These detachments give signal communications, transportation, medical, and postal services. They operate mess facilities and utilities.
- Marshal personnel from the units or from follow-on units of the ABNAF can assist in operating the marshalling areas if it does not interfere with their preparations for the airborne operation. Do not use equipment from these units because it must be packed and loaded for movement to the objective area.
- Maintain smalls stocks of supplies and equipment of all services at each marshalling area to fill last-minute shortages of the units being marshaled. Furnish services maintenance support as required.

Note. The number of personnel required to support operation and maintenance of marshalling areas varies. Based on experience, about 10 percent of the number of personnel being marshaled is required for supporting services.

FACILITY REQUIREMENTS

7-17. Commanders can use this information as a guide to selecting and modifying facilities for ABN IBCT use. Figure 7-1 shows a typical marshalling area layout for an ABN IBCT-size unit that needs about 100 acres.

Marshalling Plan

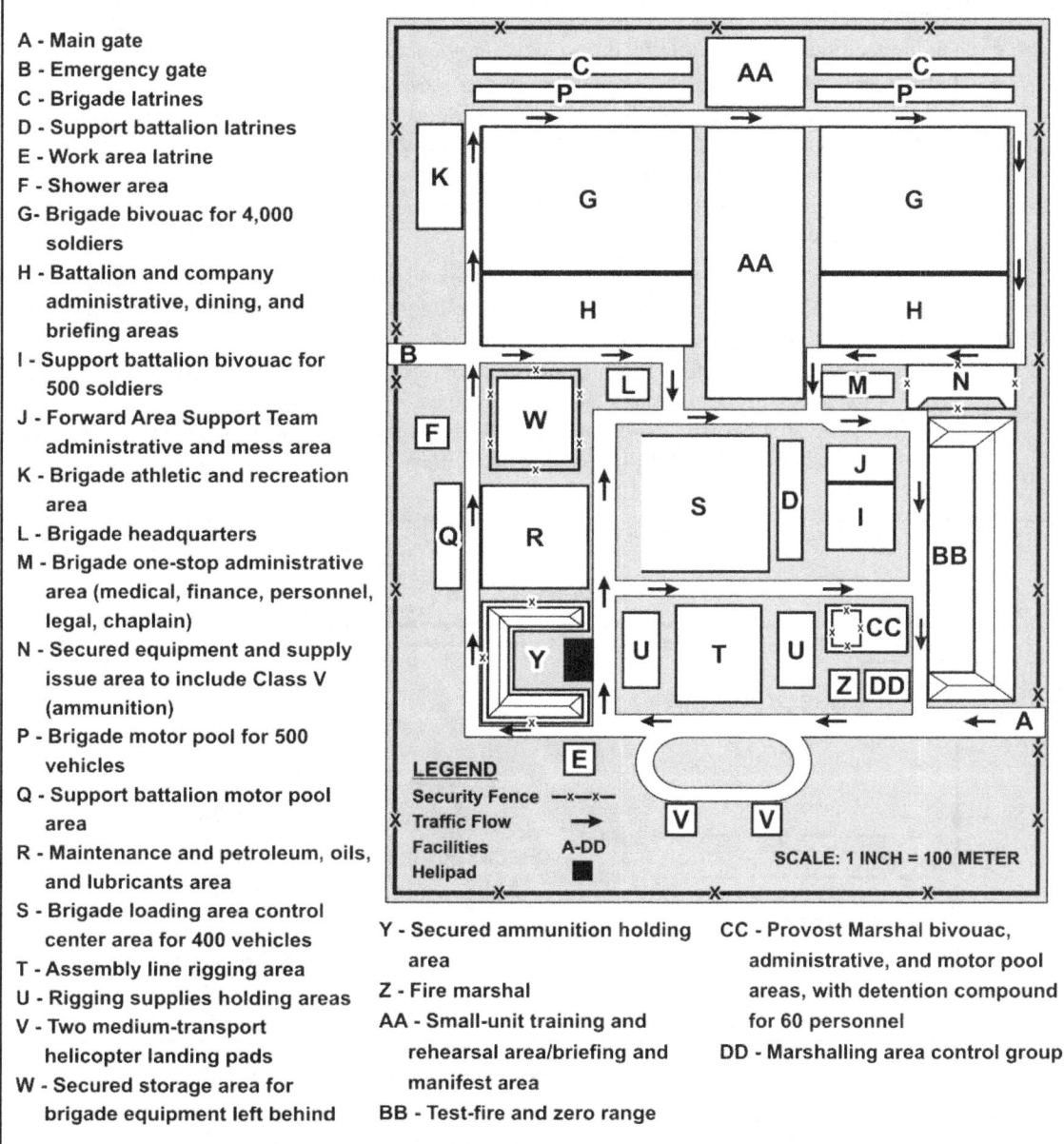

Figure 7-1. Airborne task force marshalling area

7-18. **Task force facility.** The ABN IBCT or battalion (task-organized for the mission), and the marshalling area control group occupy an ABNTF facility. If no facilities exist, support elements must construct the facilities.

7-19. **Facility specifications.** The ABNTF facility should be near a departure airfield and large enough to support the ABN IBCT or battalion (as specified), its attachments, supporting personnel, and the marshalling area control group that supports the marshalling requirements of the ABN IBCT. Each facility has a site for rigging the heavy-drop and palleted equipment for air delivery (airdrop and airland).

7-20. **Facility security.** The facility should be surrounded by security fencing or, at least, triple-strand concertina. It should have a posted security area outside the perimeter that is at least 50 meters wide and cleared of brush and trees. If available, use lights to illuminate the security area. Gates should be well lighted into the facility and have two lanes to accommodate heavy traffic.

7-21. Quarters, unit headquarters, dining areas, supply rooms, and latrines should be constructed and allocated to maintain unit integrity.

Chapter 7

- Bivouac site. If billets are not available, a bivouac site can be prepared with tents laid out in company streets.
- Dining facilities. Determining facility size requirement to adequately feed force.
- Latrine areas. There should be enough latrines to serve at least four percent of the male Soldiers and six percent of the expected female Soldiers. Build latrines at least 100 yards downwind from food service facilities to prevent food and water contamination. They need to be 30 yards from the end of the unit area, but within a reasonable distance for easy access.
- Shower facilities. Enough shower facilities should be provided to support the size force in the marshalling area.

7-22. The airborne force requires facilities for rigging heavy equipment drops and container delivery system platform loads. Although equipment can be rigged outdoors, it should be rigged in a large building, such as a hangar, where it is protected from weather. The following facilities are needed to outload:

- Loading area control center. The loading area control center (LACC) is provided for preparing vehicles for heavy-drop, or airland. It should have a 10-foot by 20-foot area for each vehicle and a 20-foot-wide area between rows for maintenance. A large area must be provided on either side of the LACC for maneuverability within the LACC for maintenance or other vehicles. (See figure 7-2.)

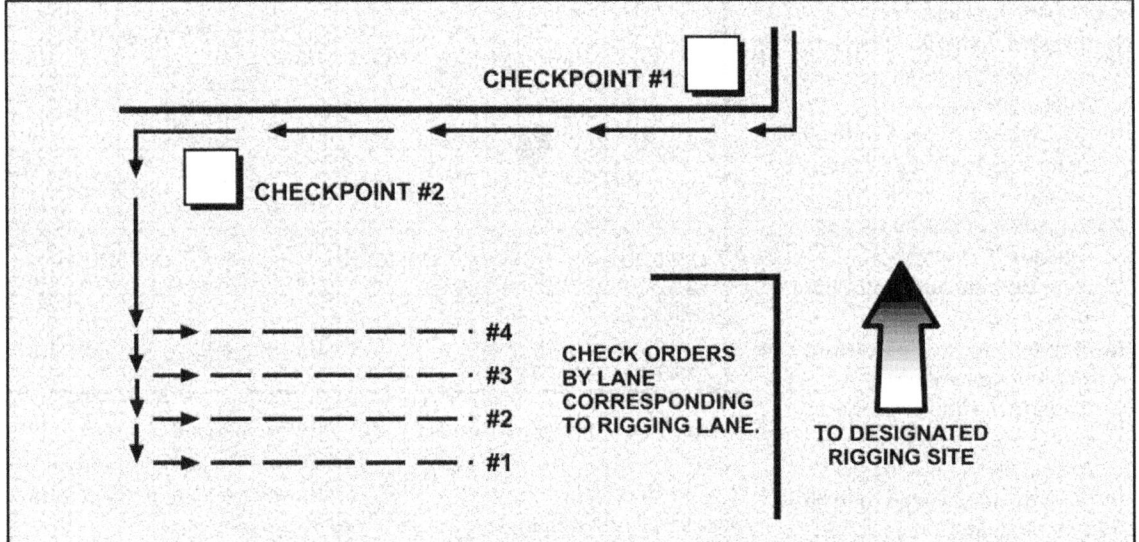

Figure 7-2. Heavy-drop loading area control center

- Rigging sites. The rigging site accommodates the rigging and outloading of about 50 platforms in a 24-hour period, depending on the availability of trained personnel, equipment, and supplies. The rigging site uses an assembly line rigging method. Riggers can operate as many lanes as required with augmentation and as available space allows. (See figure 7-3.)

Marshalling Plan

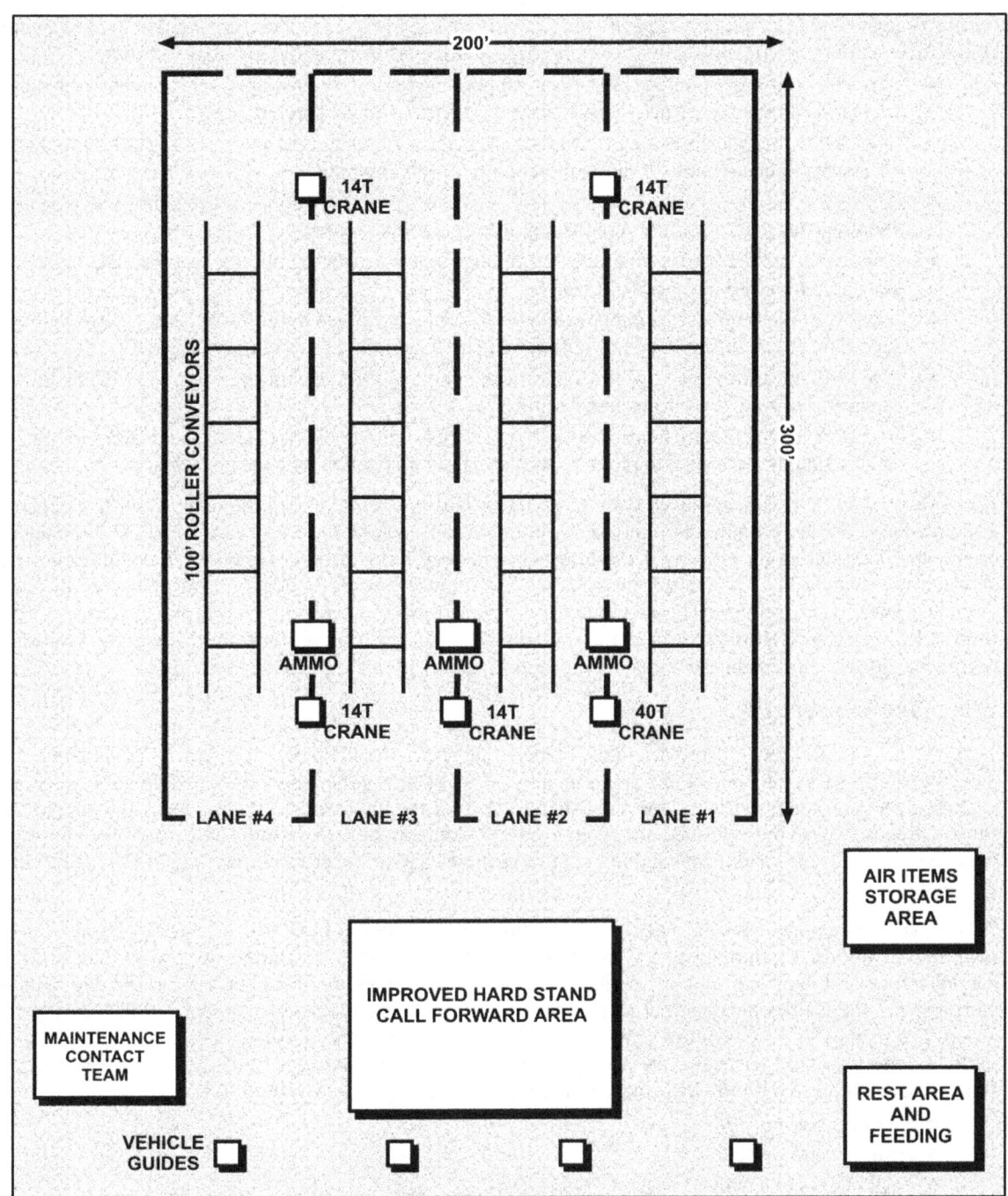

Figure 7-3. Heavy-drop rigging site

MARSHALLED UNIT AND SUPPORT ORGANIZATION ACTIVITIES

7-23. Marshalling is comprised of four activities: assemble personnel and cargo; conduct unit inspection, load equipment, and prepare; sequence loads; and establish support organization at the port of debarkation. Marshalled unit and support organization activities although not inclusive are addressed in the following paragraphs.

MARSHALLED UNIT

7-24. Prior to marshalling, if possible the marshalled unit advises the marshalling area control group, through a liaison officer or by personal contact, of the requirements for the deploying unit at the marshalling area. The marshalling area control group is the provisional unit, made up of nonorganic and certain organic units not participating in the airborne assault, organized to support the deploying unit. During marshalling coordinating staffs of the marshalled unit perform specific duties as follows:

- The S-1 provides human resources support including requisitions replacements, requests recreational facilities, legal services, and coordinates medical support.
- The S-2 is responsible for intelligence readiness to include support to security programs, counterintelligence, and deception measures to ensure secrecy.
- The S-3 submits personnel rosters, and outlines training, briefing, movement, and rehearsal requirements and tracks and submits required operational reports.
- The S-4 continues to refine the deploying unit's requirements for supply, maintenance, transportation, and storage facilities.
- The S-6 determines communications systems requirements, establishes network and electromagnetic spectrum operations, network defense, and information protection.
- The S-9 integrates civil affairs operations, gaining efficiencies presenting coordinated and deconflicted activities during marshalling.
- The staff makes requirements known as far as possible in advance of the marshalling period to enable support personnel to procure the facilities and install them where necessary.

7-25. Marshalled unit, in accordance with an unit standard operating procedure may establish a departure airfield control officer, (commonly referred to as the DACO), who will act as the ABN IBCT commander's principal representative for all activities conducted at the departure airfield. He may act as a liaison officer between the marshaled unit and the marshalling area control group or departure airfield control group, discussed later in this section. The departure airfield control officer also may control a departure airfield control team designated to perform duties as prescribed in an airborne standard operating procedures document. (Refer to unit standard operating procedures document for specific instructions.)

SUPPORT ORGANIZATIONS

7-26. When the ABN IBCT deploys and the marshalling areas close, the division support command acts as the provisional logistical unit at the home station. The theater commander responsible for the area of operation provides the provisional logistical support unit for the intermediate staging base. If a support unit cannot preposition at the intermediate staging base, a support unit from the home station command is included in the advance party. Marshalling control agencies assist the airborne and airlift force in executing the operation.

7-27. Marshalling area control group. To enable most of the airborne force to concentrate on preparing for planned operations, support agencies are designated by division headquarters to provide most of the administrative and logistical support. As stated earlier these nonorganic units and certain organic units not participating in the airborne assault organized into a provisional unit known as the marshalling area control group. The marshalling area control group, commander is the principal logistical operator for the deploying force; he executes the logistical plan. This control group provides services until the assault force departs and the marshalling area is closed. Typical assistance provided by this unit includes—

- Transportation.
- Movement control.
- All classes of supply.
- Communications.
- Facility construction, operation, and maintenance.
- Maintenance.
- Rigging.
- Recreation and other morale services.
- Local security personnel, when required.
- Army health system support.

7-28. Airlift control element (ALCE). The ALCE coordinates and maintains operational control of all airlift aircraft while they are on the ground at the designated airfield. This includes aircraft and load-movement control and reporting, communications, loading and off-loading teams, aeromedical activities,

and coordination with interested agencies, The ALCE's support function includes activities that relate to the airfield. Typical tasks for this USAF unit include—
- On both planned and rapid notice, support and control exercises and contingency operations, as defined in air mobility, and deployment and redeployment operations manuals and mission directives.
- Conduct around-the-clock operations to provide supervisory control and to ensure effective use of the airlift force on assigned missions.
- Direct, execute, and coordinate mission directives, plans, and orders assigned.
- Distribute completed loading manifests as required.
- Give copies of the aircraft-parking plan to support units.
- Coordinate loading of aircraft.
- Coordinate disposition of Army equipment and personnel remaining behind or returning because of aborted sorties.
- Ensure that appropriate and adequate briefings for Army and USAF personnel are conducted.
- Coordinate flight clearances.
- Coordinate configuration of aircraft.
- Schedule and coordinate proper air force coverage of assault landing zones, and drop zones.
- Schedule and publish air movement tables for supported units.
- Provide or arrange weather support for the mission.

7-29. Departure airfield control group (DACG). The DACG ensures that Army units and their supplies and equipment are moved from the marshalling area and loaded according to the air movement plan. Timing is critical at this point in the operation. Maintain strict control of both air and ground traffic on and across active runways.

7-30. Arrival airfield control group (AACG). Organizing the AACG is similar to the DACG. When personnel, supplies, and equipment are arriving on aircraft and need to be moved to marshalling facilities or holding areas, the AACG is responsible for offloading them. Like the DACG, the AACG works closely with the ALCE unit at the arrival airfield.

SECTION V – OUTLOAD

7-31. Complex outload operations are more difficult because they usually are conducted at night under blackout conditions. Since most or all airborne units' vehicles are rigged for air delivery, airborne forces must rely on the supporting unit for transportation during outload. These requirements are closely related to and dictated by the loading plans developed for the operation.

OUTLOAD PLANNING CONSIDERATIONS

7-32. Loading preparations are included in the marshalling plan. Loading plans outline the moving personnel and equipment and heavy-drop loads from the alert holding area to planeside. They outline the use of available materials-handling equipment. The loading plans are coordinated closely with the supporting airlift force.

7-33. A loading plan is formulated at joint conferences. It contains information about the number of personnel and the amount of equipment to be airlifted, ACLs, and the general sequence of movement.

7-34. Strict adherence to the loading timetable is mandatory. The loading of equipment and supplies must be completed in time to permit post-loading inspection, joint pre-takeoff briefing, and personnel loading by the designated station time.

7-35. The general delineations of loading responsibilities in connection with the airborne operation are as follows:
- Airlift commander responsibilities include—
  - Develops plans for specific loads and the sequence of movement in conjunction with the unit being moved.

Chapter 7

- Establishes and disseminates manifesting all cargo and personnel.
- Provides instructions for loading, instructions for documenting and unloading aircraft and for cargo tie-down.
- Parks aircraft according to the parking plan.
- Provides loading ramps, floor conveyors, tie-downs, load spreaders, and other auxiliary equipment such as operation ejection equipment.
- Prepares aircraft for ejecting cargo and for the safe exit of paratroopers from aircraft in flight. Cargo to be ejected in flight is tied down by USAF personnel.
- Ensures that a USAF representative is present to provide technical assistance and to supervise the loading unit during the loading operations of each aircraft.
- Verifies documentation of personnel and equipment.
- Furnishes and operates materials-handling equipment required in aircraft loading and unloading if the U.S. Army unit needs it.
- Airborne commander responsibilities include—
  - Establishes the priority and sequence for movement of airborne personnel, equipment, and supplies.
  - Prepares cargo for airdrop, airland, or extraction according to applicable safety instructions.
  - Marks each item of equipment to show its weight and cubage and, when appropriate, to show the center of gravity. Ensures hazardous cargo is properly annotated on DD Form 1387-2.
  - Documents and manifests all loads of U.S. Army personnel.
  - Directs and monitors both movement of ground traffic to airfield or loading area, and accepts delivery at the destination.
  - Delivers properly rigged supplies and equipment to the aircraft according to the loading plan.
  - Loads, ties down, and unloads accompanying supplies and equipment into and from the aircraft with technical assistance from a USAF representative. Cargo to be ejected in flight is tied down and ejected by USAF personnel. (Exception is made in the case of containers of supplies and equipment that are pushed from the jump exits by paratroopers immediately before their exit from the aircraft.)
  - Ensures that U.S. Army personnel are seated aboard aircraft, are properly equipped, and have their safety belts fastened by station time.
  - Briefs and supervises U.S. Army vehicle operators to ensure that the operators thoroughly comprehend airfield vehicular traffic control measures and pertinent safety precautions before they operate vehicles around aircraft.
  - Provides vehicles and loading personnel to outload U.S. Army personnel and cargo from aborting aircraft and reload them on spare aircraft if time permits.

OUTLOAD CONTROL

7-36. A control system at arrival airfields is essential to prevent congestion and to facilitate orderly movement of cargo and personnel. Outload control includes parking, traffic control, loading, bump plan, and unloading.

PARKING

7-37. The main parking consideration is loading access. Dispersal must provide the most security possible with the least possible vulnerability and, at the same time, allow maneuvering room for loading the equipment.

CONTROLLING TRAFFIC

7-38. A traffic control system is essential to avoid congestion at loading and unloading sites. In outloading, force control is accomplished by using a call-forward system in which loads are brought into the loading area as required. The following control system outline applies to air landing facilities as well as airfields. (See figure 7-4.) The system provides a separate loading facility for personnel, heavy-drop loads, and aerial supply. The separation is essential to control loading and decrease the time required to load. The airfield control system is set up with the minimum required personnel and communications equipment, and with regard to the size of the forces being moved.

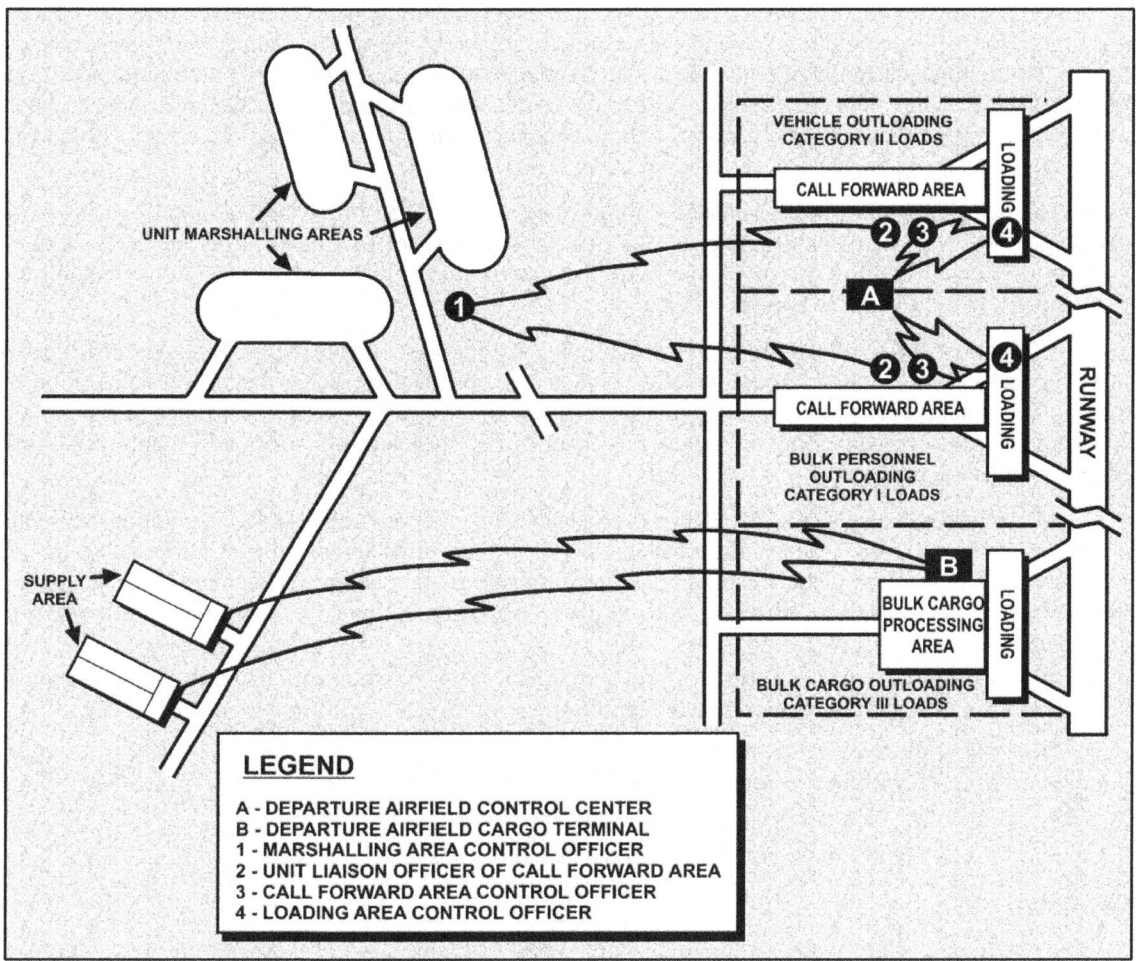

Figure 7-4. Concept of outload control

LOADING

7-39. The actual outload is complex and requires close supervision to ensure all equipment and personnel are loaded on the correct aircraft as quickly and efficiently as possible.

- Initially, personnel and equipment are dispersed in marshalling areas distant from the loading airfields, but in close communication with control groups at the airfields.
- When called, the unit or equipment is moved by planeload to the call-forward area. The fewest possible planeloads are maintained on hand in the call-forward area to ensure uninterrupted loading. Use guides and military police as required.
- As aircraft arrive in the loading area, planeloads are called forward; unit members load and tie equipment down with the technical assistance of USAF personnel.
- Control personnel maintain a log listing the departure of each aircraft. It contains the following information:
    - Aircraft tail number.
    - Summary of load or unit load number. (Correlate manifests with this entry.)
    - Time aircraft was available for loading.
    - Station time.
    - Takeoff time.

- Remarks.

BUMP PLAN

7-40. A bump plan is used to ensure that critical personnel and equipment are delivered to the airhead in the assault in the event of last-minute aircraft maintenance problems or the planned number of aircraft is not available. When designing the plan, consider communications requirements, the time required/available to execute the bump plan, and ensure secure en route communications package and tactical satellite (TACSAT) supports the physical location of key personnel (i.e. commander on right door, TACSAT on right door). A simple plan always works better under time constraints.

7-41. The bump plan must be coordinated through the ground liaison officer. State bump plan by priority and sequence, the time required to execute the plan, and number of personnel to be bumped. Units must keep in mind, the bump plan should account for the location of key personnel and communications requirements (TACSAT hatch mounts for example).

7-42. The outload brief is coordinated and facilitated by the G-3 of the higher headquarters. This is a working briefing and is the final coordination opportunity for all involved in the airborne operation. The outload coordination brief sheet is filled out during the outload brief. Special attention should be given to chalk number and formation, key personnel, key heavy drop, hot loads, air land loads, bump plan, number of personnel and time line.

7-43. All paratroopers will be marked in accordance with unit marking standard operating procedures. These markings must be strictly adhered to for both day and night jumps so that jumpmasters and key leaders can recognize the tactical cross load within a chalk, implement timely bump plans, or reorganize jumpers at the last minute to adhere to a cross load, and assemble quickly. At a minimum a bump plan should include:

- Identification and location of key personnel and communications and mission essential equipment (in accordance with the mission).
- Prioritization and sequence.
- Actions to be taken to bump non-essential personnel.
- Time required and available to execute the bump plan.
- Location of key personnel down through task force level.

UNLOADING

7-44. At arrival airfields, the control system is the reverse of that used at departure airfields. On arrival, crews unload aircraft and move the loads to dispersed holding areas where arriving elements build up to convenient size for further movements. Crews keep load categories separated to facilitate control and movement.

PART II
Air Assault Operations

Chapter 8
Organization and Employment

An *air assault operation* is an operation in which assault forces, using the mobility of rotary-wing assets and the total integration of available firepower, maneuver under the control of a ground or air maneuver commander to engage enemy forces or to seize and hold key terrain (JP 3-18). An air assault is a vertical envelopment conducted to gain a positional advantage, envelop or turn enemy forces that may or may not be in a position to oppose the operation. Ideally, the commander seeks to surprise the enemy and achieve an unopposed landing when conducting a vertical envelopment. However, the assault force must prepare for the presence of opposition. At the tactical level, vertical envelopments emphasize seizing terrain, destroying specific enemy forces, and interdicting enemy withdrawal routes.

SECTION I – AIR ASSAULT AND AIR MOVEMENTS

8-1. Air assaults are not merely movements of Soldiers, weapons, and equipment by Army aviation units and should not be considered as such. An air assault is not synonymous with an air movement. They are separate and distinct missions.

8-2. An *air assault* is the movement of friendly assault forces by rotary-wing aircraft to engage and destroy enemy forces or to seize and hold key terrain (JP 3-18). It is a precisely planned and vigorously executed combat operation. An air assault allows friendly forces to strike over extended distances and terrain barriers to attack the enemy when and where it is most vulnerable. Commanders and leaders must develop an insight into the principles governing their organization and employment to take advantage of the opportunities offered by an air assault.

8-3. *Air movement* is air transport of units, personnel, supplies, and equipment including airdrops and air landings (JP 3-17). Army air movements are operations involving the use of utility and cargo rotary-wing aircraft and operational support fixed-wing assets for other than air assaults. Air movements are conducted to move Soldiers and equipment; emplace systems; and transport ammunition, fuel, and other high-value supplies. The same general considerations that apply to air assaults apply to air movements. (Refer to FM 3-04.113 for more information.)

SECTION II – AIR ASSAULT TASK FORCE

8-4. Air assaults are accomplished by forming and employing an air assault task force (AATF). The AATF is a temporary group of integrated forces tailored to a specific mission under the command of a single headquarters. It may include some or all elements of the BCT. The ground or air maneuver commander, designated as the AATFC, commands the AATF.

ORGANIZING FORCES

8-5. The division is the lowest echelon capable of resourcing an air assault, a combat battalion or task force is the lowest echelon with a staff to properly plan an air assault, and the company is the lowest echelon with a headquarters to execute an air assault. Once the commander determines the principal components of the ground tactical plan and the maneuver and fire support schemes, the AATF organizes to execute its assigned mission. (Refer to FM 3-94 for more information.)

8-6. BCTs, combat battalions or task forces, and aviation elements from a combat aviation brigade are ideally suited to form powerful and flexible AATFs that can project combat power throughout an area of operation with little regard for terrain barriers. The unique versatility and strength of an AATF is achieved by combining the speed, agility, and firepower of rotary-wing aircraft with those of the maneuver forces in the BCTs.

BRIGADE COMBAT TEAMS

8-7. All BCTs, whether Infantry, Stryker, or Armored have the capability to plan, prepare, and execute air assault operations when the situation dictates. BCTs have maneuver, field artillery, reconnaissance and surveillance, sustainment, military intelligence, signal, and engineer capabilities that allow for the effective execution of air assault operations.

8-8. Although Stryker brigade combat teams (SBCTs) and Armored brigade combat teams (ABCTs) may not conduct air assaults as frequently as IBCTs, such operations conducted on a limited scale may be the decisive maneuver in an SBCT or ABCT operation. For this reason, all BCTs should be proficient in conducting air assaults. Examples of air assault operations conducted by SBCTs and ABCTs include seizing and retaining river-crossing sites, deliberate breach sites, and key terrain. Understanding the detailed planning and preparation that goes into an air assault enables the SBCT or ABCT to—
- Exploit the mobility and speed of task-organized or supporting helicopters to secure a key objective in the offense.
- Reinforce a threatened unit in the defense.
- Place combat power at a decisive point in an area of operation.

8-9. Due to the abundance and unrestricted use of all forms of improvised explosive devices and the threat of ground attack, it is common for aviation, engineer, and field artillery units to conduct air assault operations alongside or air movement operations separate from their Infantry counterparts. Ground tactical movement subjects the entire organization to the threat of improvised explosive devices or ground attack as soon as the vehicles leave their assembly areas or base.

COMBAT AVIATION BRIGADES

8-10. Most of the Army's aviation combat power resides in combat aviation brigades, which can be task-organized based on the mission. (See table 8-1, page 8-3, and table 8-2, page 8-4.) These combat aviation brigades include various types of organizations with manned and unmanned systems and specialize in providing combat capabilities to multiple BCTs.

8-11. In a BCT-sized air assault, the combat aviation brigade typically task-organizes based on the mission variables of METT-TC to form an aviation task force. Additional aviation companies, platoons, or sections may be task-organized to include attack reconnaissance (manned and unmanned), mission command, communications relay, air medical evacuation, and air traffic services. In BCT-sized air assaults, reinforcement with additional aviation is a common way to mass combat power and accelerate force buildup. Other combined arms forces, to include Infantry, field artillery, or engineer, and sustainment units, may be part of the combat aviation brigade task organization for specific missions.

8-12. The aviation task force employs a mix of attack reconnaissance aircraft to support an air assault, Teams of mixed attack and reconnaissance aircraft, commonly referred to as attack weapons teams and scout weapons teams, are organized and employed based on METT-TC. An attack weapons team is composed of two AH-64 attack helicopters. A scout weapons team is composed of two OH-58D scout helicopters. Planning considerations for the two airframes include:

Organization and Employment

- The AH-64 offers longer range, increased station time, and a larger weapons load, but has a larger signature.
- The OH-58D has a shorter range, decreased station time, and a lighter weapons load, but has a smaller signature.
- Both airframes have a mix of sighting systems and optics capable of employment in various environmental and weather conditions.
- Forward arming and refueling point assets are postured and configured based on airframe type.

Table 8-1. Heavy Combat Aviation Brigade organization

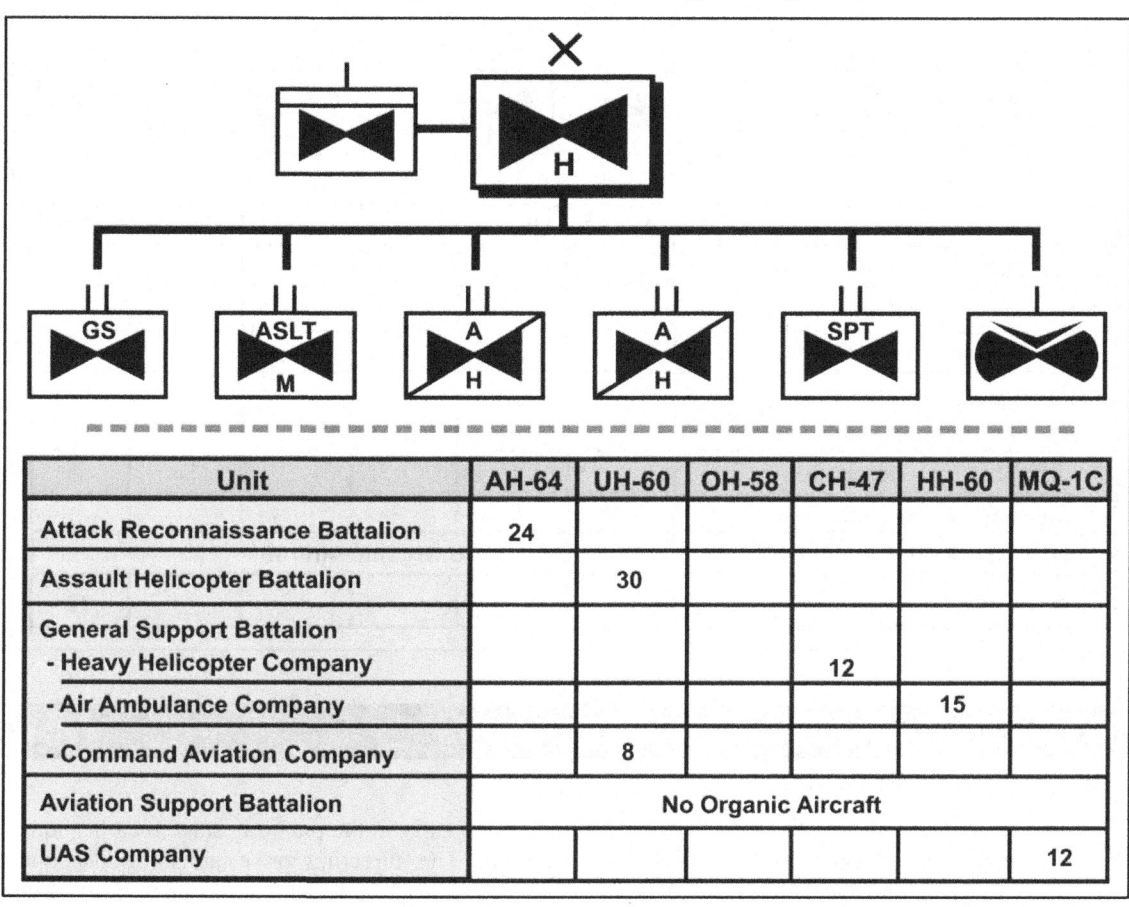

Unit	AH-64	UH-60	OH-58	CH-47	HH-60	MQ-1C
Attack Reconnaissance Battalion	24					
Assault Helicopter Battalion		30				
General Support Battalion - Heavy Helicopter Company				12		
- Air Ambulance Company					15	
- Command Aviation Company		8				
Aviation Support Battalion	No Organic Aircraft					
UAS Company						12

Chapter 8

Table 8-2. Medium Combat Aviation Brigade organization

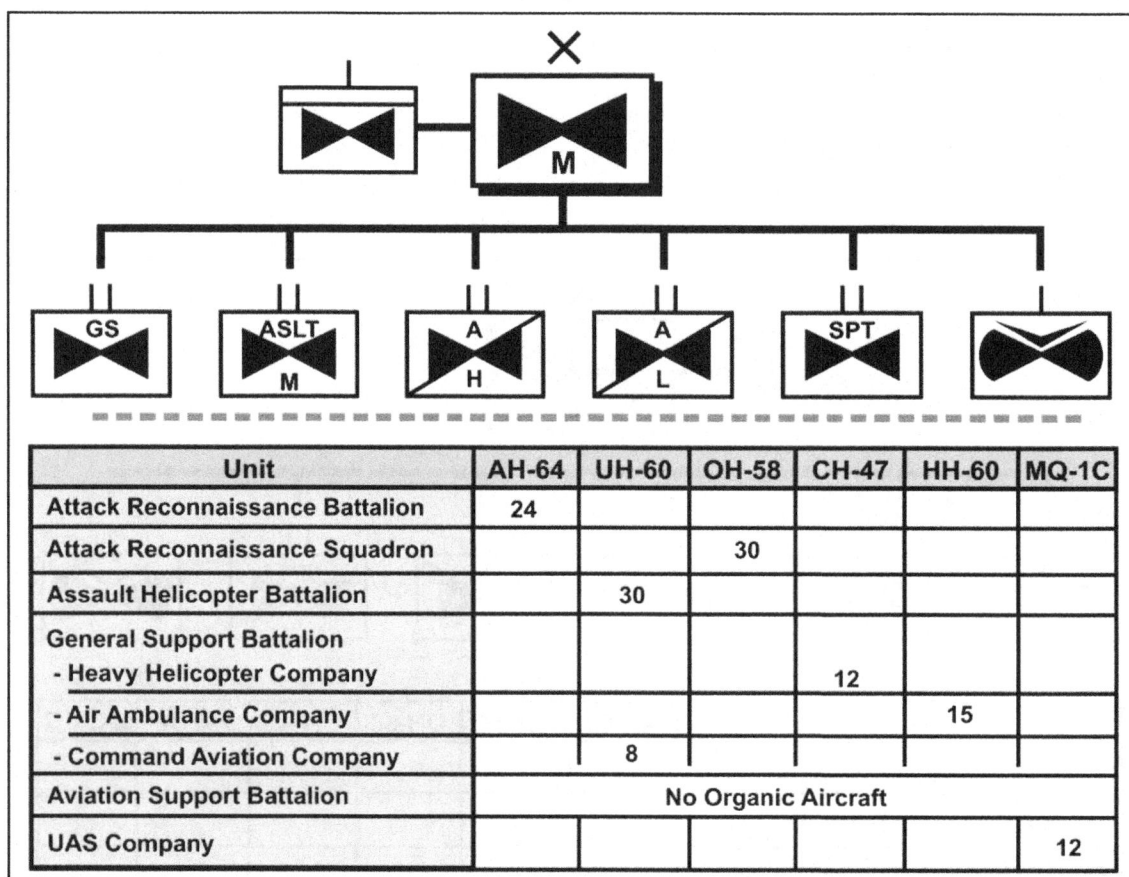

Unit	AH-64	UH-60	OH-58	CH-47	HH-60	MQ-1C
Attack Reconnaissance Battalion	24					
Attack Reconnaissance Squadron			30			
Assault Helicopter Battalion		30				
General Support Battalion						
- Heavy Helicopter Company				12		
- Air Ambulance Company					15	
- Command Aviation Company		8				
Aviation Support Battalion	No Organic Aircraft					
UAS Company						12

SECTION III – TASK FORCE CONSIDERATIONS

8-13. When forming an AATF, consider the following factors:
- Early formation of the AATF. This force is formed early in the planning stage by a headquarters that can allocate dedicated aviation resources. The directing or establishing headquarters allocates units and defines authority and responsibility by designating command and support relationships. Predesignated and well-understood command and support relationships ensure the AATF fights as a cohesive, coordinated, combined arms team.
- Availability of aviation assets. The warning order (WARNORD) may include task-organizing the AATF, which must provide a mission-specific balance of mobility and combat power. The AATF normally is organized with sufficient combat power to seize initial objectives and protect landing zones. The required combat power should be delivered to the objective area consistent with aircraft and pickup zone capacities to take advantage of surprise and shock effect.
- Maintaining unit tactical integrity. When planning loads, squads normally are loaded intact on the same helicopter, with platoons located in the same serial, to ensure unit integrity upon landing. To perform its mission, an AATF must arrive intact at the landing zone. The force must be tailored to provide en route security and protection from the pickup zone, throughout the entire air route, and at the landing zone.
- Sufficient sustainment capability. The AATF is organized with a sustainment capability to support a rapid tempo until follow-on or linkup forces arrive, or until the mission is completed. Units that support the air assault operation normally are placed in direct support to the AATF to ensure close coordination and continuous, dedicated support throughout an operation. Normally,

Organization and Employment

an AATF exists only until completing a specified mission. After that, aviation and other elements return to the control of their parent units.

SECTION IV – CAPABILITIES, LIMITATIONS, VULNERABILITIES

8-14. Gaining the initiative and setting the conditions requires commanders to know the capabilities, limitations and vulnerabilities of their force. The flexibility of conducting air assault operations allows a commander to control the tempo and exploit the initiative. A properly planned and executed air assault operation generates combat power.

8-15. An *air assault force* is a force composed primarily of ground and rotary-wing air units organized, equipped, and trained for air assault operations (JP 3-18). Air assault forces are most effective in situations where the threat of speed from mobility provides surprise and affords seizure of key terrain. Air assault forces are best suited for missions that require—

- Massing or shifting combat power quickly.
- Surprise.
- Flexibility, mobility, and speed.
- Gaining and maintaining the initiative.

8-16. Considerations for conducting air assault operations are as follows:

- Best conducted at night or during weather conditions that allow aircraft operation but obscure enemy observation to facilitate deception and surprise.
- Close air support planning must provide suppression of threats en route to and the vicinity of the vertical envelopment point.
- As early as possible, ground and air reconnaissance should be conducted at landing zones.

CAPABILITIES

8-17. An AATF can extend the battlefield, move, and rapidly concentrate combat power like no other available forces. AATF capabilities are as follows:

- Attack enemy positions from any direction.
- Conduct attacks and raids within the operational area.
- Conduct exploitation and pursuit operations.
- Overfly and bypass enemy positions, barriers, and obstacles and strike objectives in otherwise inaccessible areas.
- Provide responsive reserves, allowing commanders to commit a larger portion of his force to action.
- React rapidly to tactical opportunities, necessities, and threats in unassigned areas.
- Rapidly place forces at tactically decisive points in the area of operation.
- Conduct fast-paced operations over extended distances.
- Conduct and support deception with false insertions.
- Rapidly reinforce committed units.
- Rapidly secure and defend key terrain (such as crossing sites, road junctions, and bridges) or key objectives.
- Delay a much larger force without becoming decisively engaged.

LIMITATIONS

8-18. An AATF relies on helicopter support throughout an air assault operation. As such, it may be limited by—

- Adverse weather; extreme heat and cold; and other environmental conditions (such as blowing snow and sand) that limit flight operations, helicopter lifting capability, or altitude and elevation restrictions that affect operational capabilities.
- Reliance on air lines of communication.
- Threat aircraft, air defense, and electronic warfare action.

- Reduced ground mobility once inserted (particularly SBCT and ABCT forces).
- Availability of suitable landing zones and pickup zones due to mountainous, urban, jungle, or other complex terrain.
- Availability of air routes (for example, air routes near international borders).
- Availability of chemical, biological, radiological, and nuclear protection and decontamination capability.
- Battlefield obscuration that limits helicopter flight.
- High fuel and ammunition consumption rates.
- Availability of organic fires, sustainment assets, and protection.

VULNERABILITIES

8-19. An AATF uses helicopters to move to and close with the enemy. Initial assault elements should be light and mobile. They often are separated from weapon systems, equipment, and materiel that provide protection and survivability on the battlefield. An AATF is particularly vulnerable to—

- Enemy attack by aircraft and air defense weapon systems during the movement phase.
- Enemy attack by chemical, biological, radiological, and nuclear weapons because of limited protection and decontamination capability.
- Enemy attack by ground, air, or artillery during the loading and landing phases.
- Enemy air strikes due to limited availability of air defense weapon systems.
- Enemy electronic attack to include jamming of communications and navigation systems, and disrupting aircraft survivability equipment.
- Enemy small-arms fire that presents a large threat to helicopters during the air movement and landing phases.

SECTION V – AIRSPACE CONTROL

8-20. Airspace control requires both a control authority and a control system. Airspace control is a process used to increase operational effectiveness by promoting the safe, efficient, and flexible use of airspace with minimum restraint upon airspace users. Planning must include coordinating, integrating, and regulating airspace to increase operational effectiveness. Proper planning causes effective airspace control and reduces the risk of fratricide, enhances air defense, and permits flexibility.

CONCURRENT EMPLOYMENT

8-21. Airspace control is applying airspace control to coordinate airspace users for concurrent employment in assigned missions. Effective airspace control enables all warfighting functions to work efficiently while synchronizing air operations to support the commander's intent. Successful airspace control is dependent on the ability to perform the functions of identification, coordination, integration, and regulation of airspace users.

8-22. Properly managed airspace increases combat effectiveness. Ensuring the safe, efficient and flexible use of airspace, minimizes restraint placed on airspace users. It includes coordinating, integrating, and regulating airspace to increase operational effectiveness. Effective airspace control reduces the risk of fratricide, enhances air defense, and permits flexibility.

8-23. Airspace control does not denote ownership of a block of airspace or command over activities within that airspace. Rather, it refers to users of the airspace. All air missions are subject to the airspace control order published by the airspace control authority, which provides direction to deconflict, coordinate, and integrate the use of airspace within the operational area.

8-24. Joint forces use airspace to conduct air operations, deliver fires, employ air defense measures, and conduct intelligence, surveillance, and reconnaissance operations. At times, these missions may be time sensitive and avoid the ability to conduct detailed coordination with the land force. It is imperative that land forces provide their higher headquarters with all airspace coordinating measures to provide visibility to other joint users and prevent fratricide.

8-25. Methods of airspace control are as follows:
- Positive control relies on positive identification, tracking, and directing aircraft within the airspace control area. It uses electronic means such as radar; sensors; identification, friend or foe systems; selective identification feature capabilities; digital data links; and other elements of the intelligence system and mission command network structures.
- Procedural control relies on combining mutually agreed and promulgated orders and procedures. These may include comprehensive air defense identification procedures and rules of engagement, aircraft identification maneuvers, fire support coordination measures, and airspace coordinating measures. Service, joint, and multinational capabilities and requirements determine which method, or which elements of each method, that airspace control plans and systems use. Procedural control is a common method used by all airspace users (to include indirect fire units) to deconflict airspace. In Army rotary-wing operations, such as air assaults, procedural control is used more often than positive control.

AIRSPACE COORDINATING MEASURES

8-26. *Airspace coordinating measures* are measures employed to facilitate the efficient use of airspace to accomplish missions and simultaneously provide safeguards for friendly forces. (Refer to FM 3-52 for more information.)

8-27. Common airspace coordinating measures used during an air assault are as follows:
- Coordinating altitudes use altitude to separate users and as the transition between different airspace coordinating entities. The airspace coordinating entities should be included in the air control plan and promulgated in the airspace control order. Army echelons incorporate airspace control planning guidance and integrate the airspace control order, area air defense plan, special instructions, and air tasking order via operation orders (OPORDs). All airspace users should coordinate with the appropriate airspace coordinating entities when transitioning through or firing through the coordinating altitude.
- Restricted operations areas are airspaces of defined dimensions created in response to specific operational situations or requirements within which the operation of one or more airspace users is restricted. They are known as restricted operations zones. The AATF may use a restricted operations area or restricted operations zone to deconflict an area where prior coordination enhances aviation safety.
- Using standard Army aircraft flight routes that are routes established below the coordinating altitude to facilitate the movement of Army aviation assets. They normally are located in the corps through BCT support areas and do not require approval of the airspace control authority. They normally are listed on the current airspace control order. Direction of travel can be dictated as one- or two-way traffic.
- Using air corridors that are restricted air routes of travel specified for use by friendly aircraft and established for preventing friendly aircraft from being fired on by friendly forces. They are used to route aviation combat elements between such areas as forward arming and refueling points, holding areas, and battle positions. Altitudes of an air corridor do not exceed the established coordinating altitude.
- Using axis of advance that is a general route of advance, assigned for the purposes of control, which extends toward the enemy. The axis of advance symbol graphically portrays a commander's intention, such as avoiding built-up areas or known enemy air defense sites. When used for attack aviation operations, it provides the general direction of movement and may be subdivided into routes.
- Using air control points which are points easily identifiable on the terrain or an electronic navigational aid used to provide necessary control during air movement. Designate air control points at each point where the air route or air corridor makes a definite change in any direction and at any other point deemed necessary for timing or control of the operation.
- Using communication checkpoints which are points along the air route where serial commanders report to the air mission commander. Radio transmissions should be used only when necessary. If a report is required, consider using codes to ensure a short transmission.

Chapter 8

AIRSPACE DEVELOPMENT

8-28. When developing a course of action, the ground maneuver unit should plan an air axis of advance. This provides the general concept to the aviation planners who further refine it into routes with enough guidance to determine the direction from which the commander wants to approach. Do not submit the developed axis of advance to the higher headquarters airspace control element.

AIR ROUTE PLANNING

8-29. Upon receipt of the course of action, the aviation liaison officer plans the air routes within the air axis of advance. The aviation unit normally plans multiple routes within the axis of advance since the threat air defense disposition may not be clear. The air defense airspace management/brigade aviation element (ADAM/BAE) should assist in route planning, but the supporting aviation unit is responsible for completing the routes and submitting them to the higher headquarters airspace control element for inclusion on the airspace control order.

RESTRICTED OPERATIONS ZONE PLANNING

8-30. Any unit with organic unmanned aircraft system (UAS) is responsible for planning their own restricted operations zones for unmanned aircraft launch and recovery. All elements operating UAS in a BCT submit their request through the ADAM/BAE for deconfliction before submitting it to the higher headquarters airspace control element.

8-31. UAS launch and recovery restricted operations zones typically should be three kilometers in radius or surface to coordinating altitude, but may be tailored to meet operational requirements. Due to their size, unmanned aircraft launch and recovery restricted operations zones should not be planned near indirect fire units, supporting aviation unit assembly areas, or forward arming and refueling points (FARPs) if possible.

8-32. The supporting combat aviation brigade submits restricted operations zone locations for mission command and air medical evacuation aircraft to the higher headquarters airspace control element. Mission command and air medical evacuation aircraft restricted operation zones should be at least three by three kilometers in size. The combat aviation brigade plans both a primary and alternate restricted operations zones for each aircraft. This enables control of the operation as it moves forward and provides a restricted operations zone, if needed, for the higher headquarters mobile command group. Fire support units can utilize restricted operations zones to assist in deconflicting airspace between firing locations and target locations.

8-33. Special consideration should be given to planned employment of an organic UAS near a landing zone. If required due to the tactical mission, small-UAS should be clearly separated by a defined terrain feature from the landing zone area, and the approach and exit routes of aircraft.

METHODS TO DECONFLICT ON THE OBJECTIVE AREA

8-34. Three possible methods to deconflict airspace between attack reconnaissance aircraft and assault aircraft on the objective are described in the following paragraphs. They include: grid line or terrain feature separation, attack by fire positioning, and call clear methods.

Grid Line or Terrain Feature Separation

8-35. This is the most restrictive but easiest method to execute. It may not allow the attack reconnaissance units to engage targets in the close combat attack role during the air assault, but this technique is appropriate when time is limited for rehearsals, or when prior planning is extremely limited or not possible. With this method, the attack reconnaissance units clear the airspace for inbound assault units by moving to a designated grid line or terrain feature on either side of the objective. This movement and the subsequent maneuver of the attack reconnaissance units in and around the objective area are executed according to the instructions in the order. (See figure 8-1.)

Organization and Employment

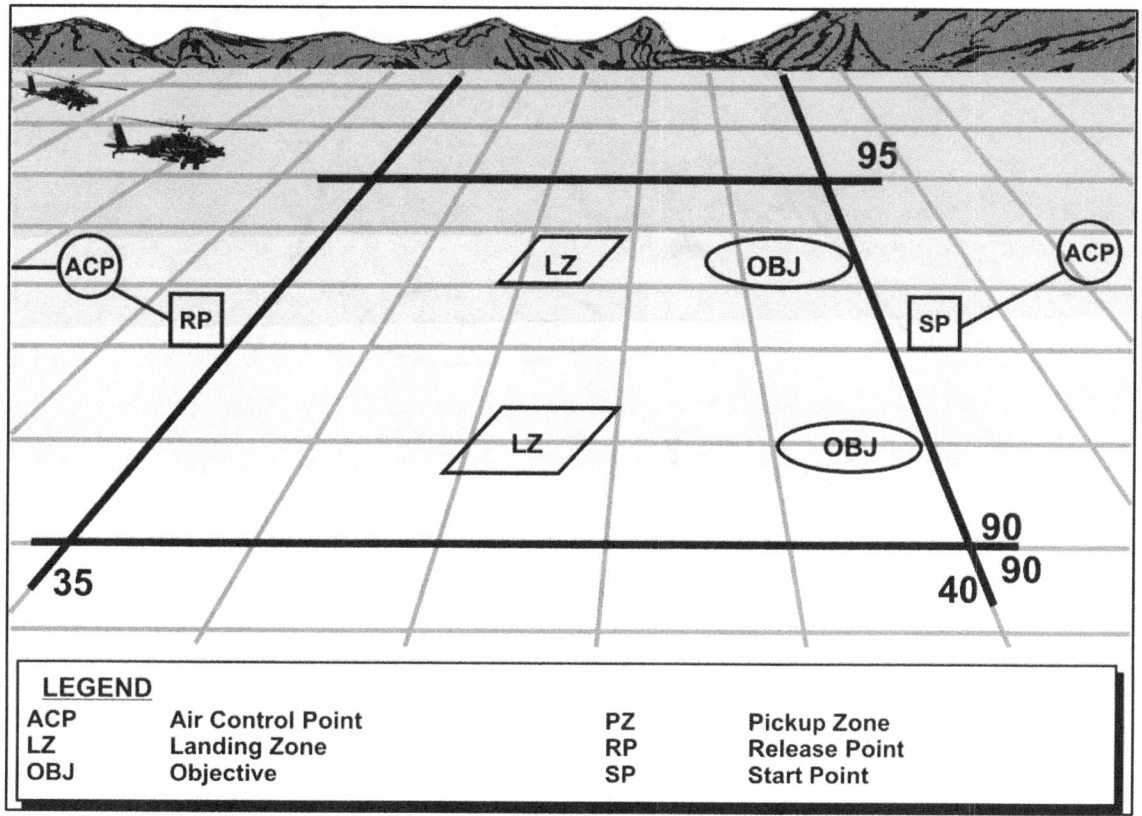

Figure 8-1. Grid line method

Attack by Fire Positioning

8-36. Attack by fire positioning is a method of deconfliction, as it allows attack reconnaissance aircraft the flexibility to engage targets during the air assault in support of the ground commander. The attack reconnaissance units occupy known attack by fire positions according to the published OPORD. This method restricts the attack reconnaissance units to the general vicinity of the attack by fire positions but not to a specific grid. (See figure 8-2, page 8-10.)

8-37. The attack by fire positioning method requires the attack reconnaissance units to ensure they stay clear of the landing zone and do not cross the centerline of the direction of flight. Using this method requires the attack reconnaissance aircraft to have increased situational awareness. This method is best used when all elements have adequate time to rehearse.

Chapter 8

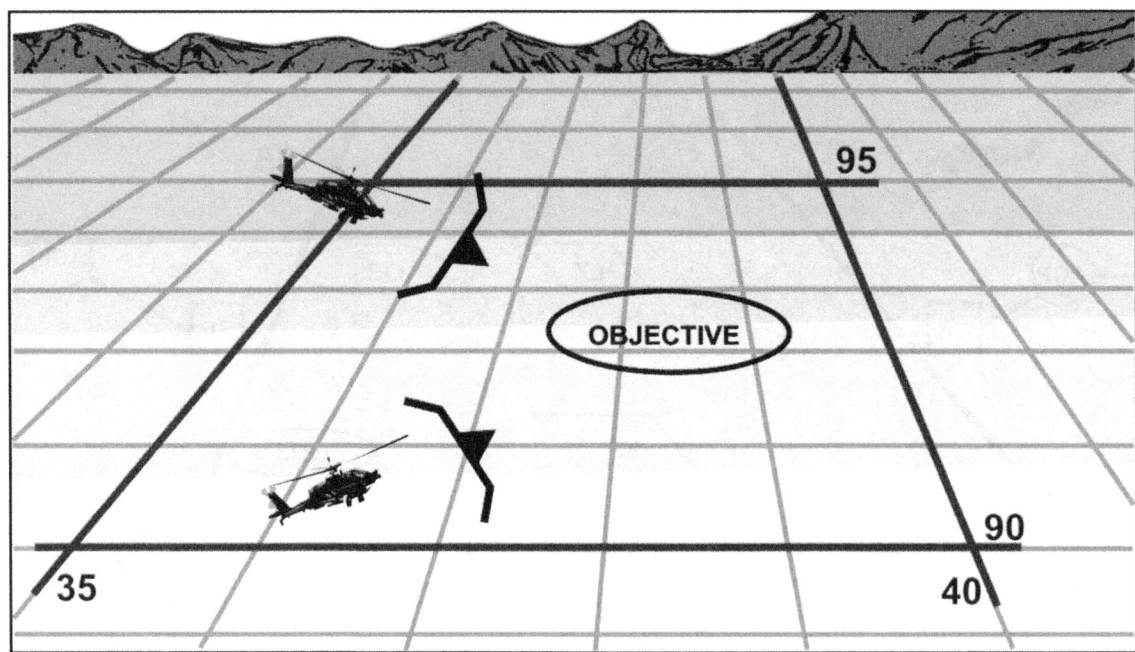

Figure 8-2. Attack by fire method

Call Clear

8-38. The call clear method is used in contingency circumstances when assault or other aircraft (such as air medical evacuation or mission command aircraft) are inbound to the objective area. It is initiated with an inbound call of the assault or other aircraft to the landing zone and a response from the attack reconnaissance air mission commander indicating that all elements of the landing zone and the flight path to it from the release point are clear. Avoid using this method during the main air assault itself due to congestion on the air battle network.

SECTION VI – AIR ASSAULT TASK FORCE MISSION COMMAND

8-39. *Mission command* is the exercise of authority and direction by the commander using mission orders to enable disciplined initiative within the commander's intent to empower agile and adaptive leaders in the conduct of unified land operations (ADP 6-0). Mission command— as a warfighting function— assist the AATFC in balancing the art of command with the science of control, while emphasizing the human aspects of mission command. Mission command systems within AATF includes the arrangement of personnel, networks, information systems, processes and procedures, and facilities and equipment that enable the AATFC to conduct air assault operations.

MISSION ORDERS

8-40. The AATFC, supported by his staff, conducts air assault operations through centralized planning and decentralized execution based on mission orders. The commander during the development of mission orders applies the foundation of mission command together with the mission command warfighting function, guided by the following principles:

- Build cohesive teams through mutual trust.
- Create shared understanding.
- Provide a clear commander's intent.
- Exercise disciplined initiative.
- Use mission orders.
- Accept prudent risk.

8-41. The AATFC's intent, formalized in the order and understood at the execution level, provides the AATF with the concept of operations (CONOPS), allowing the task force to act promptly as the situation requires. The commander focuses his order on the purpose of tasks and the air assault operation as a whole rather than on the details of how to perform assigned tasks. Orders and plans are as brief and simple as possible. (Refer to FM 6-0 for more information.)

8-42. As the commander develops his CONOPS he considers the complexity of the operation, the mission variables of METT-TC, and the experience level of his subordinate commanders and staffs to determine the detail of command. In most situations, air assaults are centrally planned and well-rehearsed before execution. This ensures that each subordinate leader knows the commander's intent and is able to execute his mission with minimal direction.

8-43. Contingencies or alternative courses of actions should be factored into the plan to allow for continuation of the mission in a dynamic environment. Tasks must be planned to occur based on time or the execution of a previous task (or tasks) so that actions occur at the specified time or in the specified sequence. Use manned or UASs for communications relay to help mitigate potentially degraded or lost communications. Commanders must plan contingencies for degraded or intermittent communications. Refer to FM 6-02 for more information.)

8-44. Another factor for the AATFC to consider when determining mission command responsibility is the location of the key AATF leadership. Key leaders should be positioned into discrete elements and dispersed throughout the lifts with provisions to ensure continuity of command. (Figure 8-3 on page 8-12 shows an example for positioning key leaders during an air assault.)

8-45. While air assault planning is centralized, air assault execution is aggressive and decentralized. Subordinate commanders should be given as much freedom of action as possible (consistent with risk, the situation, and mission accomplishment) to empower leaders to seize, retain, and exploit the initiative.

Chapter 8

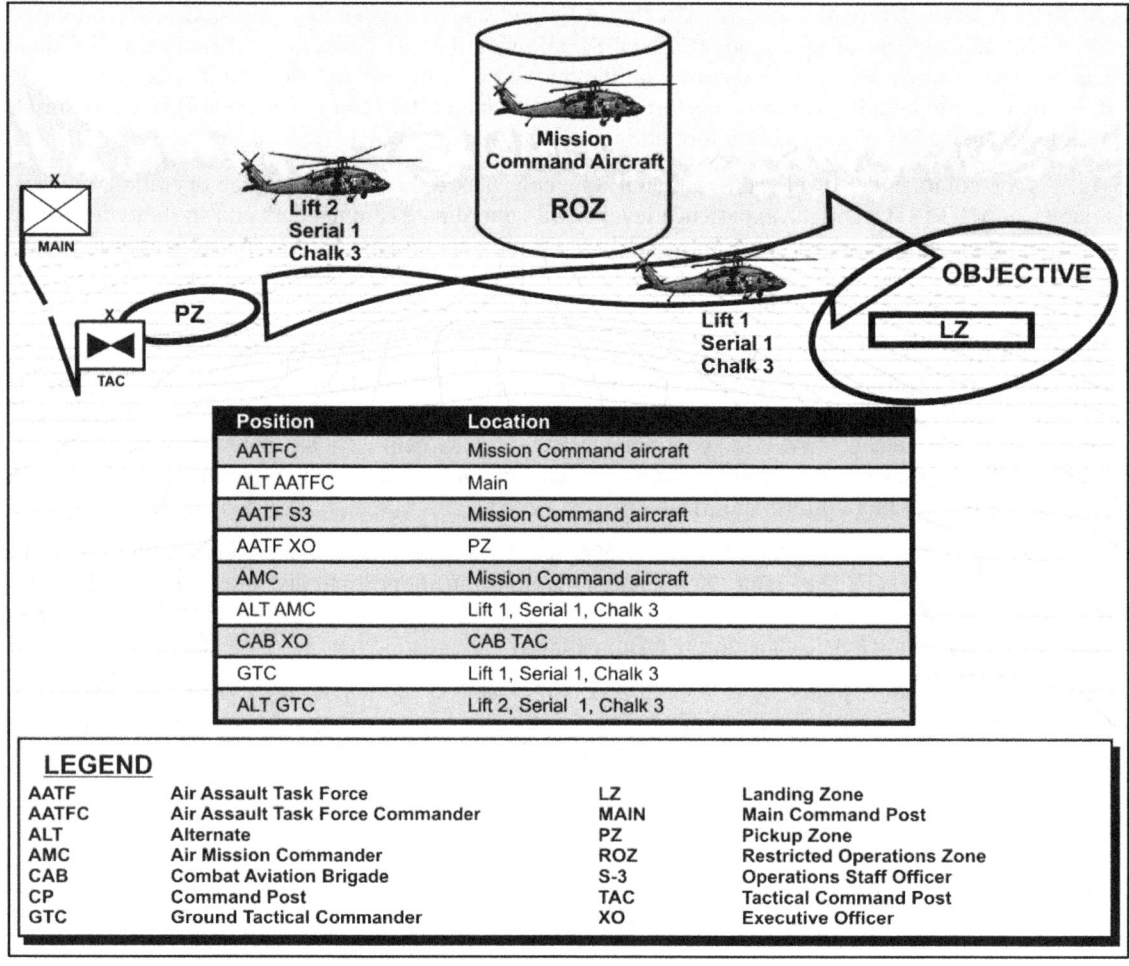

Figure 8-3. Example of air assault leadership positioning

COMMAND POSTS

8-46. The AATFC executes mission command through the establishment of two primary command posts—main command post and tactical command post. If the AATFC is the BCT commander, he has the option to form a command group consisting of select staff members who accompany him and help exercise mission command away from a command post. The following paragraphs addresses how the AATFC organizes the AATF command posts for the conduct of an air assault.

MAIN COMMAND POST

8-47. The main command post provides control of operations when the tactical command post is not deployed. When the tactical command post is deployed, the main command post—
- Provides planning for future operations.
- Maintains current enemy and friendly situations.
- Gathers information and disseminates intelligence.
- Keeps higher and adjacent organizations informed of the friendly situation and submits recurring reports.
- Acts as liaison to higher and adjacent organizations.
- Coordinates for and advises the commander on the use of enablers for future operations.
- Assists the tactical command post with executing operations as needed.
- Develops and disseminates orders as necessary.

TACTICAL COMMAND POST

8-48. The AATFC employs the tactical command post as an extension of the main command post to help control execution of the air assault for a limited period. The AATF tactical command post may be employed into the objective area soon after the initial echelon if the enemy situation permits. The tactical command post assists the commander in controlling current operations by taking the following actions:
- Maintaining the common operational picture and assisting in developing situational understanding.
- Developing combat intelligence of immediate interest to the commander.
- Maneuvering forces.
- Controlling and coordinating fires.
- Coordinating with adjacent units and forward air defense elements.
- Serving as the main command post if the main command post is destroyed or unable to function.

8-49. The tactical command post comprises the AATFC, representatives from the S-2 and S-3 sections, fire support officer, brigade aviation officer, and air liaison officer or whomever the commander designates. The tactical command post deploys in a mission command aircraft in which the air mission commander will be located. This aircraft contains a mission command package, which allows the commander to observe and direct the air assault from a forward position if he chooses. The AATFC may elect to deploy a tactical command post with the maneuver force. This command post is led by the AATF S-3 and comprises a mission-tailored portion of the AATF headquarters.

COMMAND GROUPS

8-50. *Command group* is the commander and selected staff members who assist the commander in controlling operations away from a command post (FM 6-0). The BCT headquarters can form two command groups, which are organized based on the mission. Both are equipped to operate separately from the tactical command post or main command post. Command groups give the commander and the executive officer (if required) the mobility and protection to move throughout the area of operation and to observe and direct BCT operations from forward positions.

8-51. Both command groups require a dedicated security element, additional considerations are required if ground movement is planned. The command group led by the BCT commander comprises whomever he designates. This can include the command sergeant major and representatives from the S-2, S-3, and fires sections. The commander positions his command group near the most critical event, usually with or near the decisive operation.

8-52. A second command group led by the brigade executive officer, if used, may include representation from the operations staff section, intelligence staff section, and fire support element. The executive officer usually positions his command group with a shaping operation or at a location designated by the BCT commander. The executive officer must be able to communicate with the BCT commander, battalion and squadron commanders, and command posts.

PERSONNEL AND KEY ELEMENTS WITHIN THE TASK FORCE

8-53. Emphasizing the human aspects of mission command are the personnel and key elements within the AATF who operate command posts and assist the AATFC by exercising control of the air assault from the initial planning stages through execution. This section describes the duties and responsibilities of personnel and key elements within an air assault operations..

AIR ASSAULT TASK FORCE COMMANDER

8-54. The AATFC is the overall commander of the AATF. He ensures continuity of command throughout the operation. He must position himself where he can best see the battlefield and control the operation. In situations that allow, he is airborne in a mission command aircraft during the air movement stage. At other times, he may fight the battle from a tactical command post.

GROUND TACTICAL COMMANDER

8-55. The ground commander is the commander of the largest ground maneuver force inserted during an air assault. He is usually one of the AATFC's subordinate maneuver commanders (such as a battalion or company commander). He flies on one of the first serials into the objective area, maintaining communication with the AATFC during the flight.

AIR MISSION COMMANDER

8-56. The air mission commander is the aviation unit commander or his designated representative. He receives and executes the AATFC guidance and directives, and controls all aviation elements. The air mission commander ensures continuity of command for all supporting aviation units and employs attack reconnaissance helicopters and artillery along the air route, fighting the battle from pickup zone to landing zone while keeping the AATFC informed.

AIR ASSAULT TASK FORCE S-3

8-57. The AATF S-3 assists the AATFC with mission command. He normally leads the AATF tactical command post when the AATFC is airborne in a mission command aircraft.

BRIGADE AVIATION OFFICER

8-58. The brigade aviation officer advises the AATFC on all matters relating to Army aviation and, along with the AATF S-3 Air, jointly develops the detailed plans necessary to support the air assault operation. During the execution phase, he should be available to assist the AATFC or S-3 Air in coordinating the employment of aviation units.

AIR DEFENSE AIRSPACE MANAGEMENT/BRIGADE AVIATION ELEMENT

8-59. The air defense airspace management/brigade aviation element (ADAM/BAE), led by the brigade aviation officer, is a functional element residing in the BCT's main command post. This functional element continually plans for airspace use, executes near real-time control during execution, and monitors operations of airspace users. This situational understanding is critical to ensure that the BCT can react to any situation requiring immediate use of airspace, such as immediate fires, close air support, unplanned UAS launches, or a diversion of aviation assets.

8-60. The ADAM/BAE coordinates directly with the aviation brigade or the supporting aviation task force for detailed mission planning. The ADAM/BAE element is equipped with the Tactical Airspace Integration System (TAIS), which provides a digitized, integrated, and automated system to provide airspace control and air traffic services. Shared functions between the ADAM and BAE include analysis of airspace use to determine and resolve conflicts; maintaining, requesting, and dissemination of joint airspace coordinating measures; and development and coordination of the airspace control appendix.

Air Defense Airspace Management

8-61. The ADAM plans and synchronizes air and missile defense operations with the ground commander's scheme of maneuver. The ADAM produces the integrated air picture, plans low-level sensor employment, and develops and maintains the air defense artillery overlay to include unit locations; weapons control status and weapon system coverage.

Brigade Aviation Element

8-62. The BAE advises the AATF on all tactical matters relating to Army aviation. The element provides subject matter expertise on enemy threat weapons and tactics, techniques, and procedures; aircraft survivability equipment; and mission planning and must be capable of 24-hour operations. The BAE:
- Plans and synchronizes aviation with the BCT commander's concept of operation.
- Advises and plans the use of unmanned aircraft systems, reconnaissance, attack, assault, air movement, sustainment, and medical evacuation.
- Standardizes BCT unmanned aircraft system employment.

FIRE SUPPORT OFFICER

8-63. The AATF fire support officer plans, coordinates, and synchronizes fire support for all phases of the air assault. He deploys with the AATFC in a mission command helicopter to ensure the fire support plan is executed as planned.

AVIATION LIAISON OFFICER

8-64. Although the ADAM/BAE conducts many of the functions traditionally performed by liaison officers, the aviation liaison officer from the supporting aviation brigade remains a critical part of the air assault planning process. The aviation liaison officer can be the supporting aviation unit S-3, the aviation mission survivability officer, or another aviation subject matter expert designated by the supporting aviation unit commander.

8-65. While the members of the ADAM/BAE work directly for the BCT commander as permanent staff members, aviation liaison officers represent the supporting aviation task force at a designated maneuver headquarters only for a specific operation. If colocated with the ADAM/BAE, the liaison officer team normally reports to the brigade aviation officer as a functioning addition to the ADAM/BAE staff section. Often, the aviation liaison officer coordinates with the ADAM/BAE and then proceeds to a supported ground maneuver battalion.

AIR LIAISON OFFICER

8-66. The air liaison officer (ALO) is an USAF officer who leads the tactical air control party (TACP) colocated at the BCT headquarters and advises the BCT commander and staff on air operations. The ALO leverages the expertise of the BCT TACP with links to the higher headquarters TACP to plan, coordinate, synchronize, and execute air support operations. He maintains situational awareness of the total air support and air support effects picture. Additional responsibilities of the air liaison officer include—

- Monitoring the execution of the air tasking order.
- Advising the commander and staff about the employment of air assets.
- Receiving, coordinating, planning, prioritizing, and synchronizing immediate close air support requests.
- Providing USAF input to analyses and plans.

PICKUP ZONE CONTROL OFFICER

8-67. A pickup zone control officer (PZCO) is designated for each pickup zone in an air assault. The PZCO organizes, controls, and coordinates operations in the pickup zone. Depending on the unit that is conducting the air assault, the PZCO may be a BCT, battalion, or company executive officer; BCT or battalion S-3 Air; or sometimes a company first sergeant. The PZCO operates on the combat aviation network and is prepared to assist in executing changes as needed.

MISSION COMMAND SYSTEM

8-68. The mission command system enables mission command. As with any operation the BCT uses networks and information systems, such as Brigade Combat Team Network and/or LandWarNet, to share the common operational picture with subordinates to guide the exercise of initiative. The common operational picture conveys the BCT commander's perspective and facilitates subordinates' situational understanding. This section identifies core Army battle command systems and discusses mission command systems that enable centralized planning and decentralized execution specifically to air assault operations.

INFORMATION SYSTEMS

8-69. An *information system* consists of equipment that collect, process, store, display, and disseminate information. This includes computers— hardware and software— and communications, as well as policies and procedures for their use (ADP 6-0). The Army Battle Command System gives the BCT advantages in collecting technical information, and distributing information and intelligence rapidly. The battle command

Chapter 8

system comprises core battlefield automated systems plus common services and network management. Each system provides access and the passing of information from a horizontally integrated BCT mission command network. The following are the core systems:

- Tactical Battle Command. The Tactical Battle Command System comprises the functions previously performed by the Maneuver Control System and the Command Post of the Future.
- Global Command and Control System-Army.
- Distributed Common Ground System-Army.
- Battle Command Sustainment and Support System.
- Air and Missile Defense Planning and Control System.
- Advanced Field Artillery Tactical Data System.
- Force XXI Battle Command-Brigade and Below/Blue Force Tracker.
- Tactical Airspace Integration System.
- Digital Topographic Support System.
- Integrated System Control.

INTEGRATED SYSTEM CONTROL

8-70. The BCT and battalion S-6 signal officers and S-6 sections are the air assault staff proponents responsible for planning and coordinating communications support for each phase of the air assault operation. They use integrated system control to provide communications system network management, control, planning, and support to the AATF. Also known as, the tactical internet management system, integrated system control provides network initialization, local area network management services, and an automated system to support the combat network radio-based wide area network. Features of integrated system control include mission plan management, network planning and engineering, frequency spectrum management, tactical packet network management, and wide area network management.

8-71. As the AATF executes the mission and distances become extended, communications for mission command become less sophisticated. The AATF must make extensive use of airborne or unattended very high frequency (VHF) retransmission, high frequency (HF) capabilities, and ultrahigh frequency (UHF) tactical satellite (TACSAT). Subordinate elements in the AATF may range beyond multichannel capabilities and radio transmissions, and transmissions may be unintelligible due to enemy electronic countermeasures. As a result, subordinate commanders of the AATF may have to make decisions without being in contact with the AATFC.

RADIO NETWORK

8-72. The AATF uses combat network radios primarily for voice mission command transmission and secondarily for data transmission where other data capabilities do not exist. Combat network radios are designed primarily around the single-channel ground and airborne radio system, the single-channel TACSAT, and the HF radio. (Refer to FM 6-02.53 for more information.)

8-73. AATF S-6 planners organize frequency-monitoring requirements into a communications card or matrix and distribute to key leaders, command posts, and other key personnel. Using a dynamic mix of air-to-air, air-to-ground, and ground-to-ground radio networks provides the necessary responsiveness and flexibility for air assault mission command. Table 8-3 on page 8-18 depicts the radio networks commonly employed during air assaults and recommended monitoring requirements for each. Apply the following:

- Air assault task force command network is a VHF command network dedicated to ground-to-ground coordination during operations. It normally is secure and used by the AATFC to communicate with his subordinate commanders. Given the VHF communication range limitations in restrictive terrain, consider alternate means of communications such as UHF TACSAT or HF when planning an air assault.
- Combat aviation network (CAN) is a VHF network dedicated to air-to-ground coordination during operations. All aviation elements and the remainder of the AATF elements monitor this network before and during air movements. The two combat air networks typically employed during an air assault are as follows:

Organization and Employment

- ■ CAN 1. CAN 1 provides common communications between the air assault task force commander, air mission commander, ground commander, and the pickup zone control officer.
- ■ CAN 2. CAN 2 is usually reserved as an anti-jamming network. The pickup zone control officer can use this network to provide terminal guidance to individual flight leads when required.
- Air battle network is typically a UHF command network dedicated to air-to-air communications between the air mission commander and all aviation element leaders. All aviation elements monitor this network and receive instructions from the air mission commander or the air assault task force commander when he is airborne. This network is operated on the lift unit's UHF command frequency if a dedicated airborne is not specified in the OPORD or air mission brief.
- Fire support network is a VHF network operated by the air assault task force fire support coordinator. All aviation element s must have access to this network to facilitate calls for fire during movements, insertions, and extractions. An artillery quick-fire network is used when a supporting battery or battalion is dedicated to an air assault. Plan alternate means of communication, such as TACSAT, multi-use internet relay chat, and blue force tracker or Force XXI Battle Command-Brigade and Below (FBCB2), in case of VHF communication failure.
- Operations and intelligence network is a secure VHF network controlled by the S-2 section at the main command post. All routine tactical reports and other intelligence reports are sent on this network, freeing the air assault task force commander network for command and combat critical traffic. The main command posts for all elements of the air assault task force and supporting aviation units monitor the operations and intelligence network.
- Aviation internal network is typically a VHF network operated by each aviation element leader for internal use. Using VHF radios provides each element leader with a dedicated frequency with which to direct and control individual aircraft, teams, or platoons and to communicate with air traffic control authorities.
- Pickup zone control network is a VHF network established by the pickup zone control officer for communications between ground forces at the pickup zone. The pickup zone control officer may request to use the communication platform from a mission command UH-60 if it is available. The pickup zone control officer uses this network to control the flow of vehicles in and around the pickup zone. He communicates with the pickup zone control party on this network. This ensures that chalks are lined up correctly, external loads (sling loads) are ready, the bump plan is activated if necessary, and extraneous vehicles and personnel are kept clear of pickup zone operations. All lifted units should enter the pickup zone control network 30 minutes before their pickup zone time. Specific chalks may be required to monitor the network if the aircraft formation in the pickup zone requires them.

Table 8-3. Standard air assault radio networks and monitoring requirements

	AATF Command Network	CAN 1	CAN 2	ABN	FS/Quick-Fire Network	O&I Network	AVN TF Network	PZ Control Network
AATFC	X	X		X		X		
Ground CDR	X	X			X	X		
AMC	X	X		X	X		X	
FSO		X	X	X				
AVN LNO		X		X			X	
PZCO		X	X					X
Lifted Unit	X	X	X		X	X		X

LEGEND
AATF Air Assault Task Force FS Fire Support
AATFC Air Assault Task Force Commander FSO Fire Support Officer
ABN Air Battle Network LNO Liaison Officer
AMC Air Mission Commander O&I Operations and Intelligence
AVN Aviation PZ Pickup Zone
CAN Combat Aviation Network PZCO Pickup Zone Control Officer
CDR Commander TF Task Force

Chapter 9
Air Assault Planning

Planning for air assault operations mirrors the military decisionmaking process (MDMP). It incorporates parallel and collaborative planning actions necessary to provide the additional time and detailed planning required for successful execution of the air assault mission. Standardizing operations between units conducting the air assault significantly enhances the ability of the unit to accomplish the mission.

SECTION I – ROLES AND RESPONSIBILITIES

9-1. Air assault planning is as detailed as time permits and should include completion of written orders and plans. Within time constraints, the air assault task force commander (AATFC) carefully evaluates capabilities and limitations of the total force and develops a plan that ensures a high probability of success. The planning time should abide by the one-third/two-thirds rule to ensure subordinates have enough time to plan and rehearse.

HIGHER HEADQUARTERS

9-2. The headquarters above the element forming the air assault task force (AATF) directs the formation of the AATF. This headquarters allocates units, defines authority, and assigns responsibility by designating command and support relationships. The staff of this headquarters is responsible for developing the task organization of the AATF and conducting the necessary steps of the MDMP. A division-level commander or his equivalent is the approving authority for the formation of an AATF larger than a company.

BRIGADE COMBAT TEAM

9-3. The brigade combat team (BCT) is the core of the air assault and the BCT commander is normally the AATFC for a battalion air assault. The AATFC for a company air assault is normally the battalion commander of the company conducting the assault. The primary role of the AATFC and his staff is to develop the ground tactical plan by providing his staff and the supporting aviation unit staff with key tasks, intent and aviation staff guidance concerning the weight of the attack reconnaissance coverage and the level of acceptable risk for the supporting aviation units.

SUPPORTING AVIATION UNITS

9-4. Normally, supporting aviation units are operational control to the AATFC. Occasionally, an aviation task force may be created to support an AATF. However, the combat aviation brigade commander typically anticipates the needs of the AATFC and provides the necessary aviation units to support the mission of the AATF. As the supporting unit, the combat aviation brigade commander directs aviation units within his command or requests augmentation from his higher headquarters to meet the needs of the AATFC.

SECTION II – REVERSE PLANNING SEQUENCE

9-5. Air assault planning is based on careful analysis of the mission variables of METT-TC and detailed reverse planning. Five basic plans comprise the reverse planning sequence. (See figure 9-1, page 9-2.) They are—
- Ground tactical plan.
- Landing plan.
- Air movement plan.
- Loading plan.

Chapter 9

- Staging plan.

9-6. These plans are not developed independently. The AATF staff and supporting aviation unit coordinate, develop, and refine concurrently to make best use of available time and resources. They develop the ground tactical plan first, which serves as the basis to develop the other plans. Each plan may potentially affect the others. Changes in an aspect of one plan may require adjustments in the other plans. The AATFC must determine if such adjustments entail acceptable risk. If the risk is unacceptable, the concept of operations (CONOPS) must change.

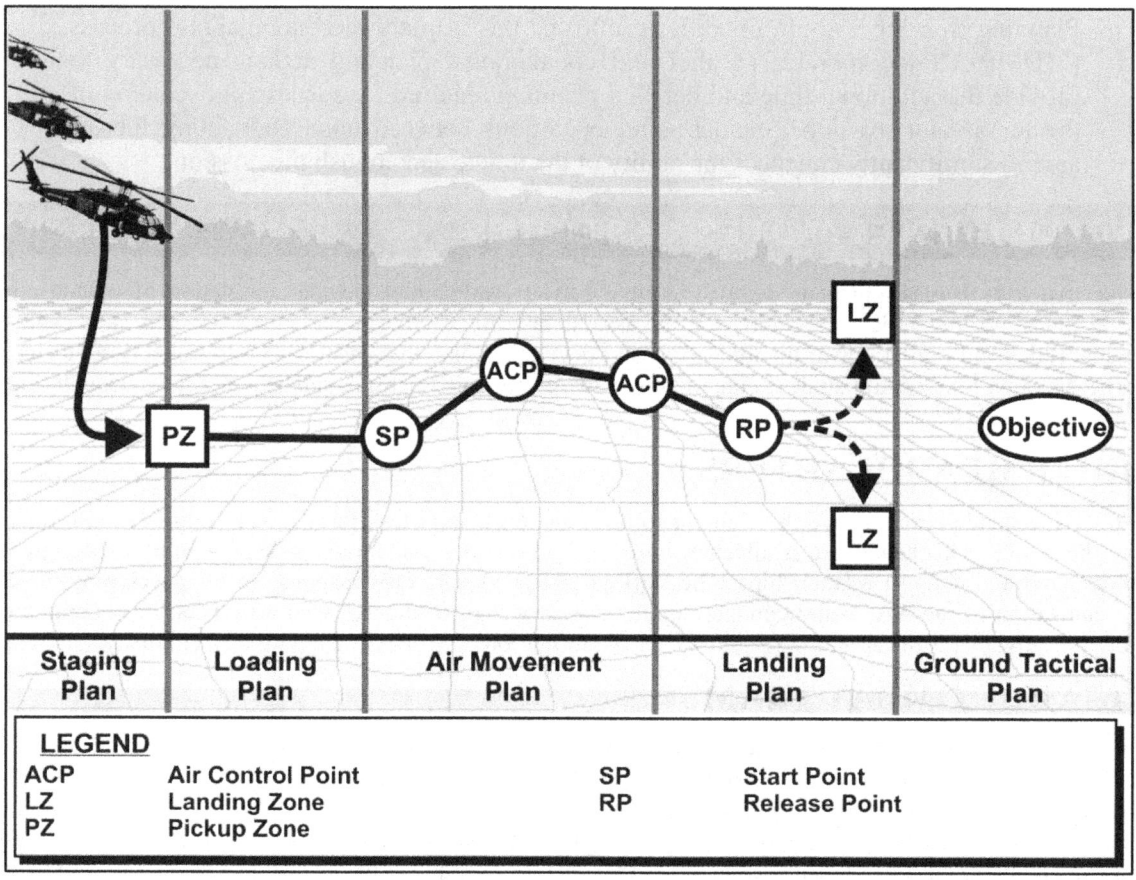

Figure 9-1. Air assault planning stages

SECTION III – PLANNING METHODOLOGY

9-7. The AATF staff conducts continuous coordination under the guidance of the AATF executive officer and S-3 during course of action development. Concurrent planning includes identifying air assault hazards and developing controls to mitigate risk.

9-8. The air defense airspace management/brigade aviation element (ADAM/BAE) and the supporting aviation liaison officer from the combat aviation brigade serve as the link between the combat aviation brigade staff and the AATF staff and are critical to the air assault planning process during mission analysis and course of action development. The ADAM/BAE and the aviation liaison officer serve as the subject matter experts on aviation operations to the AATF and supported unit staffs during this process. This enables the AATFC and ground commander to concentrate on refining the ground tactical plan and follow-on missions. The ADAM/BAE must anticipate requirements of the supported units and disseminate these requirements as soon as possible to the aviation liaison officer.

9-9. The AATF, supported unit staff, and supporting aviation unit staff should receive and share the following:

- Landing zone confirmations by imagery, aircraft videos, landing zone sketches, reconnaissance products, patrols, and higher headquarters intelligence.
- Composition of assault, follow-on, and area of operation echelons by unit.
- Nomenclature of every vehicle and sling load to be flown and maximum expected weight and air item availability for heavy and light loads.
- Confirmed troop counts by serial for assault and follow-on echelons.

9-10. The collaboration between the AATF, supported unit staff, and supporting aviation unit staff results in the Air Assault Appendix to Annex C (Operations), of the OPORD and may include—
- Tentative lift and serial composition (draft air movement table).
- List of suitable pickup zones and landing zones.
- Tentative air routes.
- Landing zone imagery (if available).
- Any deviations from standard planning factors.
- An execution checklist.

DELIBERATE PLANNING

9-11. Air assaults are deliberately planned due to the complex nature and requirement to provide the commander detailed intelligence concerning the enemy situation. The air assault planning process mirrors the steps in the MDMP and incorporates parallel actions necessary to provide the additional time and detailed planning required for successful mission execution. (Figure 9-2 on page 9-4 provides a comparison of the MDMP and the air assault planning process when maximum time is available for planning.)

Chapter 9

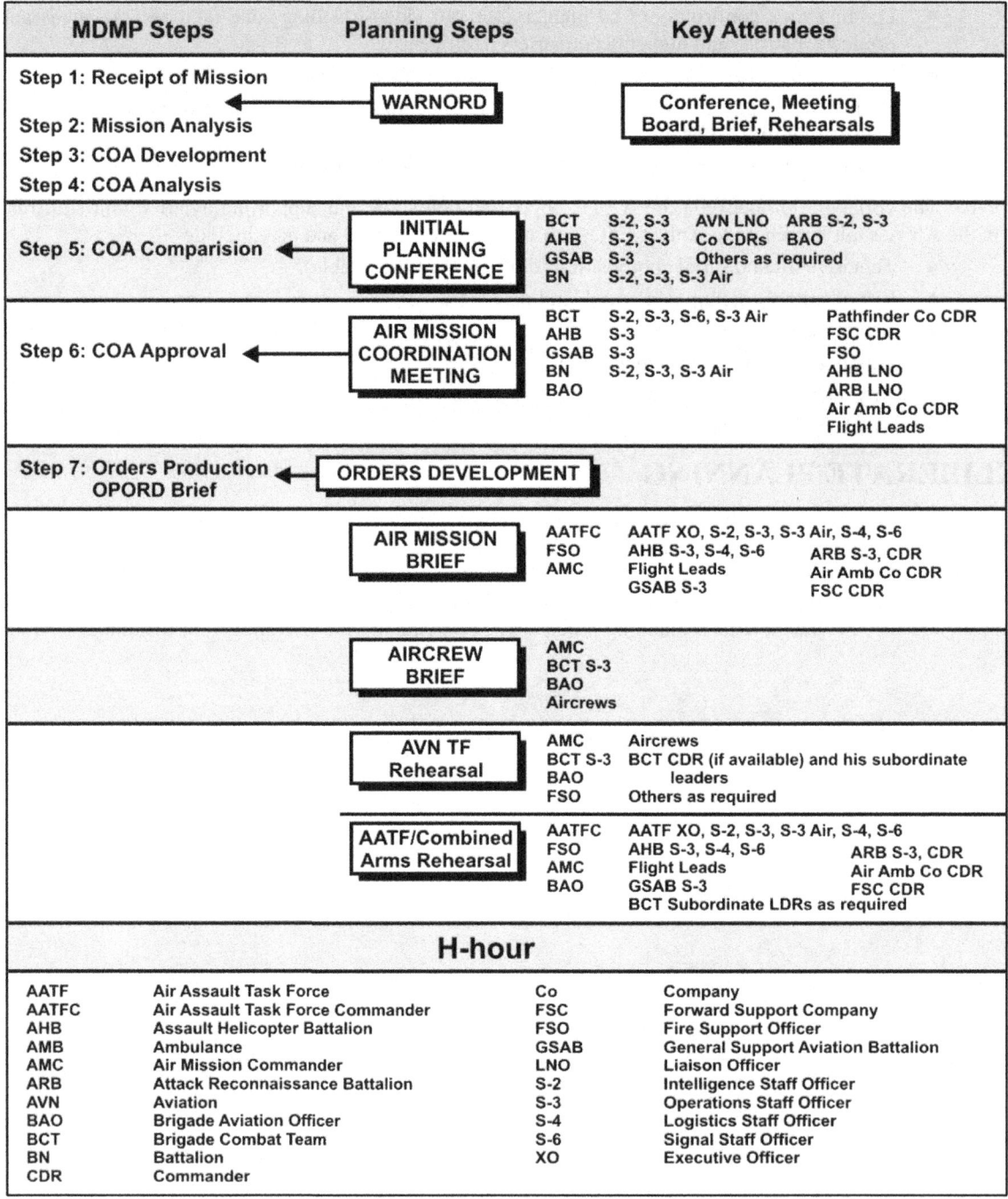

Figure 9-2. MDMP and air assault planning process

TIME-CONSTRAINED PLANNING

9-12. Due to the dynamic nature of operations, units often are required to execute air assaults within short time constraints, sometimes a few hours from the time of receiving the OPORD. Based on the time available, the AATF executive officer adjusts the timeline as required. It is critical for the executive officer to consider the ability of the supporting aviation unit to accomplish its tasks with its crew endurance program. (See figure 9-3, page 9-5.)

Air Assault Planning

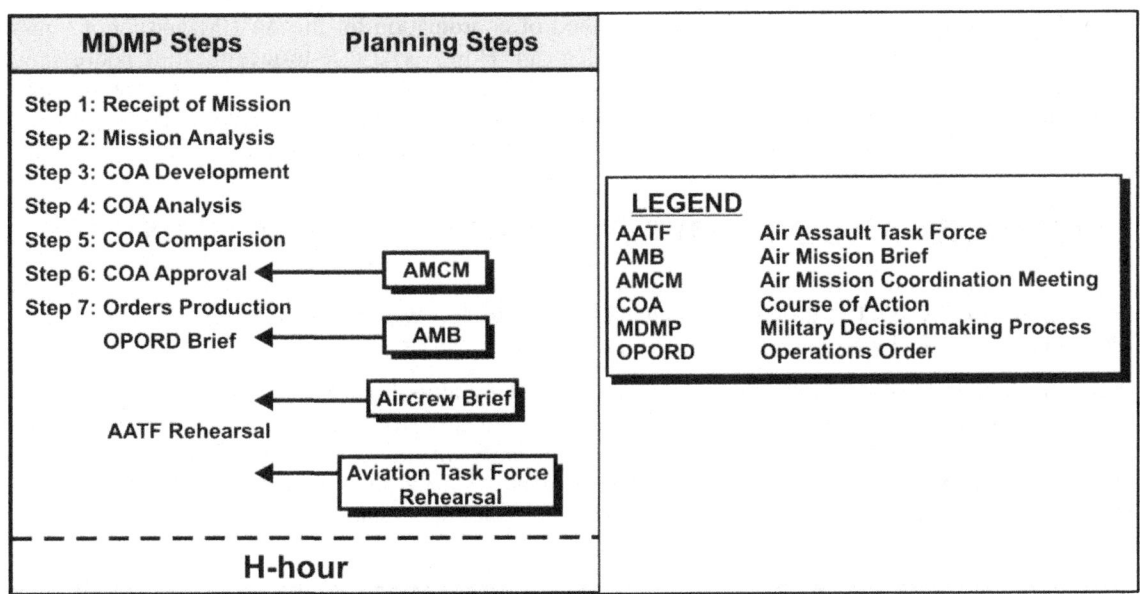

Figure 9-3. Time-constrained air assault planning

9-13. Successful execution of an air assault in a time-constrained environment requires parallel and collaborative planning by all units and staffs that are part of or supporting the AATF. Parallel planning begins as soon as the mission is received, with the supporting combat aviation brigade providing liaison officers or conducting coordination through the ADAM/BAE to the AATF if they are not colocated with unit. Through continual coordination with the supporting combat aviation brigade, the ADAM/BAE advises the AATF S-3 on limitations of aircraft or crew availability affecting course of action development.

9-14. Once the AATFC has provided a directed course of action or approved a course of action, the brigade aviation officer immediately begins the air mission coordination meeting (AMCM). To save time by reducing the number of meetings, specific portions of the air mission brief are included in the AATF OPORD brief in lieu of doing a separate air mission brief. With the exception of the combined OPORD and air mission brief, mission coordination and planning may be completed by phone, e-mail, video teleconferencing, or other mission command systems. Backbriefs, aircrew briefs, and rehearsals still are conducted as described earlier. The specific portions of the air mission brief included in the OPORD brief are—

- Staging plan.
- Air movement plan.
- Landing plan.
- Attack reconnaissance coverage.
- Landing zone condition criteria.
- Weather decision.
- Risk assessment.

RAPID DECISIONMAKING AND SYNCHRONIZATION PROCESS

9-15. Typically, the AATFC executes air assaults in response to time-sensitive intelligence or rapidly changing battlefield conditions. If time is extremely limited, the AATFC may choose to rely on his intuition and direct the staff to use the rapid decisionmaking and synchronization process (RDSP). While the MDMP seeks the optimal solution, the RDSP seeks a timely and effective solution within the commander's intent, mission, CONOPS, and level of risk. Using the RDSP lets the staffs avoid the time-consuming requirements of developing and comparing multiple courses of action. (Refer to ADRP 5-0 for more information.)

9-16. Due to a shortage of time, the primary method of coordination for the AMCM and the air mission brief may be via video teleconference or conference call. However, a face-to-face meeting addressing the contents of the air mission brief should be conducted before mission execution. This meeting may be conducted on the pickup zone with aircraft shutdown. At a minimum, the flight lead, air mission commander, chalk leaders, S-2, and the ground commander should be present. Rehearsals conducted in this situation should be combined AATF and aviation unit events.

ALLOWABLE CARGO LOAD PLANNING CONSIDERATIONS

9-17. To load an effective AATF aboard helicopters, commanders and staffs must know the exact composition of the AATF, the essential characteristics of the types of helicopters to be used for the operation, and the methods of computing aircraft requirements.

9-18. Maximum allowable cargo load (ACL) is affected by altitude and temperature and differs widely according to topography and climatic conditions common to specific zones or areas of military operations. ACLs vary based on the location of, approaches to, and exits from landing zones; pilot proficiency; aviation unit standard operating procedures; type of engine in the aircraft; and age of both aircraft and aircraft engine. Therefore, two identical aircraft, of the same model and type, may not be able to pick up and carry identical loads.

SECTION IV – PLANNING PROCESS

9-19. Throughout the operations process, the AATFCs and staffs synchronize the warfighting functions to accomplish missions. Commanders and staffs use several integrating processes and continuing activities to do this. Synchronization is the arrangement of action in time, space, and purpose, integration is combining actions into a unified whole. (Refer to ADRP 5-0 for more information.)

9-20. The integrating processes combines the efforts of the AATFC and staff to synchronize specific functions throughout the operations process. The integrating process includes—
- Intelligence preparation of the battlefield. (Refer to ATP 2-01.3 for more information.)
- Targeting. (Refer to FM 3-60 for more information.)
- Risk management. (Refer to ATP 5-19 for more information.)

9-21. The AATFC and staff ensure several continuing activities are continuously planned and coordinated. The following continuing activities require particular concern of the commander and staff throughout the operations process:
- Information collection. (Refer to FM 3-55 for more information.)
- Security operations. (Refer to FM 3-90-2 for more information.)
- Protection. (Refer to ADRP 3-37 for more information.)
- Liaison and coordination. (Refer to FM 6-0 for more information.)
- Terrain management. (Refer to ADRP 3-90 for more information.)
- Airspace control. (Refer to FM 3-52 for more information.)

9-22. MDMP integrates activities of the commander, staff, subordinate commanders, and other military and civilian partners when developing an air assault OPORD. The AATF staff fosters a shared understanding of the situation as it develops a synchronized plan or order to accomplish a mission.

WARNING ORDER

9-23. Air assault planning begins when the designated AATF receives a warning order (WARNORD) from higher headquarters for the upcoming air assault mission. The WARNORD specifies the AATFC and task organization. This allows the aviation commander to dispatch a liaison officer to the AATF headquarters early in the planning phase. Other WARNORDs and fragmentary orders (FRAGORDs) should follow as the AATF staff and commander work through the reverse planning sequence.

9-24. The following information is sent out with the WARNORD to provide units in the AATF the information needed for planning:
- Ground commander's scheme of maneuver.

- Estimate of the size of the force to be air assaulted.
- Likely pickup zones and landing zones.
- Air assault task force commander's intent on the number of lifts and general timeline.
- Initial estimate on requirements for attack reconnaissance aircraft.

INITIAL PLANNING CONFERENCE

9-25. The initial planning conference is the first meeting between the AATF staff and supporting aviation unit. The air mission commander, liaison officer, assault helicopter battalion S-2 and S-3, flight leads, and select aviation brigade staff personnel should represent the aviation unit. This initial meeting allows the supporting aviation unit planners to address impacts that environmental factors (climate and weather, terrain, and altitude) may have on the performance capabilities of the aircraft and subsequent mission accomplishment with the AATF planners, as early as possible in the planning process. The initial planning conference is conducted at the AATF headquarters.

9-26. The AATF staff should have hastily war-gamed the concept for the ground tactical plan before the initial planning conference in order for planners to discuss and determine landing zones, routes, and pickup zones. If more planning time exists, units may conduct a subsequent AMCM (similar to the initial planning conference), but this occurs after the ground tactical plan and other mission details are finalized.

9-27. Following the initial planning conference, the ground and aviation staffs should understand the distance and general time involved for each lift. The staffs should know which forces are planned to be in the first lift and in each serial of the first lift, and which first-lift serials are going to which landing zones and by what route. Subsequent lifts and follow-on echelon lifts, while discussed at the initial planning conference, can be planned in detail at a later AMCM if time permits.

AIR MISSION COORDINATION MEETING

9-28. The air mission coordination meeting (AMCM) is a meeting between the AATF and supporting aviation units. An S-3 meeting follows the development of the ground tactical plan. The AMCM is run by the brigade aviation element and chaired by the AATF S-3. The AMCM is scheduled to allow sufficient time for maneuver units to decide on a specific course of action based on the WARNORD and the standard planning factors.

9-29. The AATFC should approve the maneuver course of action before the AMCM. At the AMCM, unit S-3s brief the concept of their ground tactical plans. S-3s show the composition of combat power, by echelon, required at each landing zone. It is imperative that the subordinate unit S-3s attend this meeting with an 80 to 90 percent solution on their requirements.

9-30. The meeting is not complete until the assault helicopter liaison officers know which loads go to which landing zone and in what sequence. Attack reconnaissance liaison officers must know the air routes to be used, and all must understand the landing zones and agree on a tentative air movement table with the start and end times of the first and last serial on the landing zone. The brigade aviation element is the central figure in coordinating this information.

9-31. The AATF S-3, executive officer, or commander must approve changes after the AMCM. It is critical that the supported unit and the air assault planners come to the AMCM with the information needed for an effective meeting. (See table 9-1, page 9-8.) The end result of the AMCM is a finalized air movement plan, landing plan, air routes, pickup zones, and landing zones.

Table 9-1. Example of an air mission coordination meeting agenda

Roll Call	Brigade Aviation Officer/S-3 AIR
Intelligence Update	Air Assault Task Force S-2
Weather	Air Assault Task Force S-2
Ground Tactical Plan	Air Assault Task Force S-3
Air Movement Plan	Assault Helicopter Battalion LNO
Attack Reconnaissance Aviation Concept	Attack Reconnaissance Battalion LNO
Fires	Air Assault Task Force FSO
Mission Command	Air Assault Task Force S-6
Medical and Casualty Evacuation Plan	AATF Forward Support Medevac Team LNO/S-1/S-4
Refueling Plan	Assault Helicopter Battalion LNO
Load Plan	Brigade Aviation Officer/S-3 AIR
Review Decisions	Air Assault Task Force S-3
AATF S-3 Closing Comments	Air Assault Task Force S-3

NOTE: If reconnaissance or pathfinder insertions are planned, also cover emergency extraction plan/trigger, alternate communications plan, rehearsals, communications check, and final coordination. For artillery raids, include larger time/location and trigger for extraction.

LEGEND
AATF	Air Assault Task Force	S-2	Intelligence Staff Officer
FSO	Fire Support Officer	S-3	Operations Staff Officer
LNO	Liaison Officer	S-4	Logistics Staff Officer
		S-6	Signal Staff Officer

AIR MISSION BRIEF

9-32. Air mission brief refers to the written product and the briefing itself. The air mission brief is a coordinated staff effort during which the AATFC approves the air assault plan. The air mission brief is in addition to the AATF OPORD and is published in the air assault appendix to the operations annex. (Refer to FM 6-0 for more information.)

9-33. The air mission brief highlights air assault requirements to the AATF, aviation, and ground units. It should not be a working meeting. It is a backbrief to the AATFC and important to the key subordinate aviation and ground unit leaders who execute the mission. The combat aviation brigade or supporting aviation unit staff plays a vital role in the air mission brief process.

9-34. The air mission brief should stress assault and attack concepts, sequence of events, and the reasoning for the mission's sequence. The slightest change in serial separation, landing zones, or other elements of the mission can significantly affect the rest of the plan. The AATFC must approve changes to the air assault mission after the air mission brief. It is difficult to resynchronize the different warfighting functions in the short time that remains between the air mission brief and mission execution.

AIR MISSION BRIEF DOCUMENTS

9-35. Documents required to conduct a thorough air mission brief include—
- Air movement table. The air movement table regulates the sequence of flight operations from pickup zone to landing zone. (Refer to chapter 12 of this publication for more information.)
- Communications card. The communications card includes a summary of all call signs and networks.
- Pickup zone diagrams. The pickup zone diagram graphically depicts the pickup zone. Units should prepare a separate diagram for each light and heavy pickup zone.

- Landing zone diagrams. Graphically depicts the landing zone and should be prepare for each light and heavy landing zone.
- Operations sketch. Sketch provided by each battalion S-3 describing the ground maneuver plan and given to the aviation S-3 at the BCT rehearsal. Each pilot carries an operations kneeboard sketch to provide situational awareness and to counter the potential for fratricide during close combat operations. These sketches are included as enclosures to the air mission brief. Additional documents that enhance the operations sketch include the grid reference graphic and a concept of fires.
- Route cards. Depict ingress and egress routes on the air assault.
- Execution checklist. The air assault execution checklist permits brief, informative radio transmissions on crowded radio networks. Execution checklists will use brevity codes to represent critical points in the scheme of maneuver. Ensure brevity codes are aligned with multi-service brevity codes. (Refer to ATP 1-02.1 for more information.)

AIR ASSAULT TASK FORCE REHEARSAL

9-36. The AATF combined arms rehearsal is culminating the formal air assault planning process. It is a rehearsal of the entire air assault mission, beginning with condition setting and ending with the commander's expressed end state.

9-37. The rehearsal includes the aviation flight lead, S-3, the AATF staff, and other key leaders. The emphasis is on synchronizing all units supporting and executing the air assault. Included in the rehearsal is a discussion and demonstration of likely ground and air contingencies, such as downed aircraft, alternate route or landing zone activation, delays in the pickup zone, alternate suppression of enemy air defenses (SEAD) plan, and others suited to a particular mission.

9-38. It is critical that air assault security forces from attack reconnaissance aviation units are represented at the rehearsal to confirm air route deconfliction, fire control measures, and locations of expected attack by fire or battle positions. Additionally, the AATF S-3 or their designated representatives attend the rehearsal to brief the ground tactical and fire support plans.

AIRCREW BRIEF

9-39. In the aircrew brief, aviation unit and serial commanders brief all flight crews executing the air assault mission. The aircrew brief covers all essential flight crew actions and aviation planning necessary to accomplish a successful mission. Flight crews must fully understand the mission to execute the air assault successfully.

9-40. The aircrew brief is conducted at the aviation battalion level, with the aircrews from each unit in attendance. The aircrew brief can be conducted at the aviation company level (with assistance from the aviation brigade staff) when conditions do not allow the brief to be conducted at the battalion level.

AVIATION TASK FORCE REHEARSAL

9-41. The aviation task force combined arms rehearsal is similar to the AATF rehearsal. However, its emphasis is the aviation scheme of maneuver and the contingencies associated with the movement of aircraft and how they apply to the mission. The purpose of the aviation task force rehearsal is to validate synchronization.

9-42. At a minimum, the rehearsal includes the pilot in command of each aircraft, the air mission commander of each serial, the aviation task force S-3, and the aviation task force commander. Additional requirements are set by the air mission commander. Topics discussed should include, but are not limited to, route deconfliction, bump plan execution, execution matrix, downed aircraft recovery procedures, personnel recovery, actions on contact, and pickup zone and landing zone procedures.

CONDITION CHECKS

9-43. Condition checks are coordination meetings conducted by the AATF staff to update the AATFC on the status of how well shaping operations create the conditions to execute the air assault. These conditions are monitored constantly to ensure they exist for air assault execution. It is important to consider the latency of the information when presenting it to the commander for a decision.

9-44. The initial air assault condition check usually is conducted in the AATF or ground tactical force main command post. All air assault staff principals are represented. BCT and higher headquarter liaison officers attend each other's condition checks in person when possible and by video-teleconference or conference call when necessary. The final condition check is conducted near the AATF's pickup zone control command post. It includes a review of the latest friendly, terrain and weather, and enemy situations.

9-45. An air assault condition check considers critical factors to evaluate and recommend the execution of an air assault. For example, air assaults planned for dawn and dusk periods are extremely dependant on weather and visibility. Air assaults planned for these periods increase the risks to air assets. The S-2 evaluates the weather and visibility conditions and provides recommendations to the AATFC based on his assessment.

ABORT CRITERIA

9-46. Abort criteria is a predetermined set of circumstances, based on risk assessment, which makes the success of the operation no longer probable; thus the operation is terminated. These circumstances can relate to changes in safety, equipment or troops available, preparation or rehearsal time, weather, enemy, losses during execution, or a combination of the above. The methodology used in executing an air assault involves setting the conditions, providing suppressive fires immediately before and on landing, and continuously monitoring abort criteria from beginning to end.

CONSIDERATIONS

9-47. Abort criteria are important considerations when a change of one or more conditions in the objective area or landing zone seriously threatens mission success. As such, they are the friendly force information requirement relating to ongoing air assault operations and requiring command consideration regarding mission continuation. It is important that the air mission brief clearly defines abort criteria and that the AATFC monitors them throughout the operation.

DECISION PROCESS

9-48. If an abort criterion is met, a decision sequence is used before aborting the mission—
- Delay. If time is available, delay a mission in order to correct a circumstance that may abort a mission.
- Divert. If time is not available or a delay does not correct an abort criterion, the task force may execute a divert contingency away from its primary air assault mission.
- Abort. If an abort criterion exists and a delay or diversion to the mission does not correct it, the mission can be aborted by the AATFC. Apply the following:
  - A lift is aborted when it reaches an abort criterion. The mission itself is not aborted.
  - A mission is aborted when an abort criterion exists for the entire mission and the AATFC decides to abort.

9-49. Given the continued advantage of using the primary landing zone over the alternate, delay while en route or at the pickup zone is preferable to diverting. The AATFC must evaluate the risk of such a delay in light of time, fuel, enemy, and other mission variables.

9-50. Planners establish proposed abort criteria to assist commanders in deciding when success of the operation is no longer probable. The AATFC retains authority for abort decisions. The six factors that determine abort criteria for air assault missions are as follows:
- Weather. Adverse weather conditions make flying unsafe and degrade the effectiveness of the helicopter's organic weapon systems. The support combat aviation brigade sets theater-specific minimum weather conditions and establishes the appropriate approval authorities for risk management.
- Available aircraft. The ground tactical plan for an air assault operation depends on the rapid massing of combat power at the critical place and time by helicopters. The supporting aviation

task force manages combat power to support the AATF and keeps the staff informed of any limitations.
- Time. Refers to three distinct subjects: light and darkness, planning time, and fighter management.
  - Light and darkness. U.S. armed forces gain a significant advantage over most military forces in the world by operating at night. Night operations may increase aviation survivability, but may increase accidental risk in periods of low illumination. The aviation task force standard operating procedure specifies illumination thresholds related to mission approval.
  - Planning time. In general, less planning time equates to increased risk. Time-sensitive operations should be preplanned to the greatest extent possible and should rely on established standard operating procedures.
  - Fighter management. Aircrew fighter management may impact the air assault timeline if the mission is delayed or extended. The aviation task force standard operating procedure specifies approval level for mission extension.
- Mission essential combat power. Air assault mission planners use mission variables to determine the minimum combat power (to include Infantry, field artillery, and aviation) needed to ensure mission success. Use abort criteria to ensure friendly forces have the required combat ratio for the operation.
- Mission criticality. The success of units and future operations may depend on the success of the air assault mission. Therefore, some air assault operations may proceed despite the presence of circumstances that normally would abort the mission.
- Enemy. Certain types of enemy activity, especially along air routes or near landing zones or objectives, may abort an air assault mission. Abort criteria usually is stated in terms of the size or type of an enemy unit, the type of enemy equipment (especially air defense), and the proximity of the enemy to present or future friendly locations.

SECTION V – CONTROL MEASURES

9-51. A *control measure* is a means of regulating forces or warfighting functions (ADRP 6-0). Conduct of air assault operations is inherently complex and requires unity of command. Complexity of the operation necessitates keeping the operation as simple as possible with control measures.

FORCE-ORIENTED CONTROL MEASURES

9-52. Coordinated force-oriented control measures are crucial and allow for maneuvering ground and air elements to operate with confidence. Control measures are preplanned steps that establish boundaries of an area of operation.

9-53. Commanders use boundaries to assign units tactical responsibility of a designated geographical area. They use control measures to govern airspace control and clearance on the ground of a diversity of fire support ranging from artillery and naval gun fire (to close combat attacks and close air support. Commanders try to use easily identifiable terrain as a reference aid to enhance fire support coordination measures (FSCMs) and easily recognizable terrain features on the ground to expedite maneuver.

9-54. An important point on maneuver control graphics; is that staffs must be knowledgeable regarding the different maneuver control measures and their impact on clearance of fires. For instance, boundaries are both restrictive and permissive, while corridors, routes, and direction of attack are restrictive.

9-55. Since boundaries serve as both permissive and restrictive measures, the decision not to employ them has profound effects upon timely clearance of fires at the lowest possible level. This is important, especially if maneuver units are not given areas of operations meaning that no boundaries are established.

9-56. The higher echelon may coordinate all clearance of fires short of the coordinated fire line, a time-intensive process. It allows the unit to maneuver successfully and to engage targets in a swift and efficient manner. It requires coordination and clearance only within that organization.

Chapter 9

BOUNDARIES

9-57. A *boundary* is a line that delineates surface areas for the purpose of facilitating coordination and deconfliction of operations between adjacent units, formations, or areas (JP 3-0). (Refer to JP 3-09, FM 3-09, and FM 3-90-1 for more information.) Boundaries affect fire support in two ways. They are—

- Restrictive. Boundaries are restrictive in that normally units do not fire across boundaries unless the fires are coordinated with the adjacent unit or the fires are allowed by a permissive fire support coordination measure, such as a coordinated fire line. These restrictions apply to conventional and special munitions and their effects. When fires such as obscuration and illumination affect an adjacent unit, coordination with that unit normally is required. A commander may employ direct fires without clearance at specific point targets that are clearly and positively identified as enemy. Targets and their triggers should be kept within the same unit's boundary without overriding other tactical or doctrinal considerations.

- Permissive. Boundaries are permissive in that a maneuver commander, unless otherwise restricted, enjoys complete freedom of fire and maneuver within his own boundaries. Thus, units may execute joint fires without close coordination with neighboring units unless otherwise restricted.

FIRE SUPPORT COORDINATION MEASURES

9-58. A *fire support coordination measure* (FSCM) is a measure employed by commanders to facilitate the rapid engagement of targets and simultaneously provide safeguards for friendly forces (JP 3-0). The following restrictive and permissive FSCMs— are used frequently in airborne or air assault tasks. See JP 3-09, FM 3-09, and FM 3-90-1 for a discussion of FSCMs.

- Restrictive Fire Support Coordination Measures. Restrictive measures impose requirements for specific coordination before engagement of targets. Restrictive FSCMs are those that provide safeguards for friendly forces and noncombatants, facilities, or terrain. Restrictive FSCMs include the no-fire area, restrictive fire area, restrictive fire line, fire support area, fire support station, and the zone of fire.

- Permissive Fire Support Coordination Measures. Permissive FSCMs facilitate the attack of targets by reducing or eliminating the coordination necessary for the clearance of fires. Permissive FSCMs include the coordinated fire line, fire support coordination line, free fire area, and the kill box.

AIRSPACE COORDINATING MEASURES

9-59. *Airspace coordinating measures* (ACM) are measures employed to facilitate the efficient use of airspace to accomplish missions and simultaneously provide safeguards for friendly forces (JP 3-52). Two ACMs–the airspace coordination area and the restricted operations area– are frequently used in airborne or air assault tasks. Refer to JP 3-52, FM 3-52, FM 3-90-1 for a discussion of ACMs.

- An *airspace coordination area* is a three-dimensional block of airspace in a target area, established by the appropriate ground commander, in which friendly aircraft are reasonably safe from friendly surface fires. The airspace coordination area may be formal or informal (JP 3-09.3). The airspace coordination area is the primary ACM which reflects the coordination of airspace for use by air support and indirect fires. (Refer to JP 3-09.3 for more information.)

- A *restricted operations area* (ROA) is airspace of defined dimensions, designated by the airspace control authority, in response to specific operational situations/requirements within which the operation of one or more airspace users is restricted (JP 3-52). An ROA is used to separate and identify areas, examples include but are not limited to artillery, naval surface fire support, unmanned aircraft system operating areas, areas of combat search and rescue, special operations forces operating areas, and areas which the area air defense commander has declared weapons free. Air defense missions generally have priority over ROAs. (Refer to JP 3-52 for more information.)

Air Assault Planning

SECTION VI – SHAPING OPERATIONS

9-60. A shaping operation is an operation at an echelon that creates and preserves conditions for the success of the decisive operation. Shaping operations establish conditions for the decisive operation through effects on the enemy, population, and terrain.

9-61. The AATFC determines the exact conditions that must be created and preserved according to the mission variables of METT-TC and the level of risk associated with each air assault. When determining these conditions, the AAFTC considers the following factors:

- Shaping operations are not limited to conducting ground and air reconnaissance, suppression of enemy air defenses, and preparation fires. They may require additional augmentation from higher headquarters, the supporting aviation unit, and the supported unit staffs to succeed.
- Assessing the effects of lethal fires by conducting battle damage assessment of enemy forces and capabilities is not easy. Enemies often remove wounded or dead personnel and equipment to make friendly battle damage assessment more difficult and less accurate.
- In weighing the validity of battle damage assessment projections, it is important to balance confirmed intelligence against friendly combat power applied. More combat power may be useful against uncertain battle damage assessment.
- The threat, the ability to assess the impact of shaping operations, and the air assault execution time may determine the duration of shaping operations. Allocate as much time as possible.

9-62. Conducting shaping operations to create and preserve the proper conditions for air assault execution is an iterative process. Based on his situational understanding, the AATFC decides what part of the situation must change to ensure success of the air assault. The commander directs available reconnaissance forces and surveillance assets to detect the location of enemy systems that unacceptably endanger the air assault's success. This allows lethal and nonlethal systems, such as artillery, jammers, attack reconnaissance aircraft, and unmanned aircraft systems (UASs), to target and deliver the desired fires and/or effects against enemy systems before launch.

9-63. The AATF staffs continue to plan and prepare for the air assault. The AATFC considers employing service and joint fires to help set the conditions. The commander requests assistance from higher headquarters if sufficient organic assets and information are not available to accomplish the mission. The commander then assesses the progress of the shaping operations. This process repeats until the commander is satisfied with the result or operational necessity forces him to either abort or conduct the air assault.

9-64. When available Army teams from the long-range surveillance company (LRSC), a divisional or corps asset, are organized, trained, and equipped to deploy into the objective area and conduct reconnaissance and surveillance tasks before the deployment of the air assault force. Special operations forces may be inserted or already be operating in the objective area and become key components of the initial effort to shape and set conditions. Special operations (to include special reconnaissance missions) are keys to setting conditions, and integrated into the operation at every stage from initial planning to transition. (See FM 3-05.) Refer to chapter 1 of the publication for additional information.

9-65. The AATFC employs his reconnaissance and surveillance forces (to include cavalry, scouts, chemical, biological, radiological, and nuclear platoon, and UAS to conduct reconnaissance and surveillance of proposed landing zones and the objective area to identify and target enemy forces near the landing zones and objectives. The field artillery battalion is positioned to provide fires throughout all phases of the operation. Shaping operations should deny the enemy's ability to conduct reconnaissance, defeat his strike operations, and neutralize his ability to communicate and command. The supporting aviation unit is prepared to conduct reconnaissance in coordination with reconnaissance forces or provide lethal fires to neutralize or destroy enemy forces in the objective area once they have been identified.

9-66. The commander considers employing other capabilities that may not be in his task force, such as close air support and electronic warfare assets. The purpose of these operations is to set and preserve the conditions on the landing zone and objective area that allow the maneuver forces to launch the air assault and execute a successful attack to destroy the enemy on the objective.

9-67. Successful execution of an air assault may be decisive to accomplishing the mission, but it is not necessarily the decisive operation. Air assaults often are conducted as shaping operations to establish the conditions for the decisive operation through the effects rendered on the enemy and terrain. An example of this is a company conducting an air assault to seize a bridge and secure a crossing site in support of a combined arms battalion-level attack that requires the bridge as a crossing site.

9-68. Similarly, the AATF sets the conditions for a successful air assault by conducting shaping operations of its own. The AATFC may employ ground and air reconnaissance units, attack aviation units, UAS, close air support, and artillery fires to conduct shaping operations to mitigate the level of risk for executing the air assault.

SECTION VII – MEDICAL AND CASUALTY EVACUATION

9-69. Medical evacuation refers to both air and ground casualty evacuation. Air medical evacuation employs air assets from the air ambulance companies assigned to the combat aviation brigade and air medical evacuation general support aviation battalions to evacuate casualties. Dedicated air medical evacuation aircraft include specifically trained medical personnel to provide en route care. The nine-line medical evacuation request is the standard method to request medical evacuation.

9-70. Casualty evacuation refers to the use of nonmedical vehicles or aircraft to evacuate casualties. Use casualty evacuation only when the number of casualties exceeds the medical evacuation assets or when the urgency of evacuation exceeds the risk of waiting for medical evacuation assets to arrive. Typically, air assaults plan for both air and ground evacuations. (Refer to FM 3-04.113 and ATP 4-02.2 for more information.)

MEDICAL EVACUATION PLANNING

9-71. The combat aviation brigade allocates medical evacuation assets to the supported AATF for the duration of the air assault. However, the size and distance of the planned air assault dictates the duration of medical evacuation support to the AATFC. As a rule, the supporting commander should provide medical evacuation assets to the supported commander until ground lines of communications are established.

9-72. Typically, the evacuation platoon leader from the brigade support medical company and the air ambulance platoon leader from the air ambulance company conduct the medical evacuation planning for the air assault. They do so in coordination with the AATF S-1, AATF S-4, BAE, AATF S-3, supported unit S-3, BCT surgeon section, and BCT support medical company commander. The air ambulance platoon leader should brief the medical evacuation plan at the AMCM, AMB, and health service support rehearsal. When planning for medical evacuation during an air assault they should—

- Integrate ground evacuation measures into the overall medical evacuation plan.
- Plan medical evacuation routes to Level II or III health care facilities. Ensure all aircrews participating in the air assault know these routes.
- Plan for medical personnel to fly on casualty evacuation aircraft if time and situation permit.
- Ensure medical evacuation crews are available for air assault orders, rehearsals, and preparations.
- Brief casualty collection point locations and markings during the air assault rehearsal.
- Plan to maintain a forward arming and refueling point (FARP) after the air assault is completed so that medical evacuation aircraft have a staging place for follow-on ground tactical operations.

9-73. Medical evacuation aircraft are limited assets and should be scheduled and used accordingly. The AATF's casualty estimate provides planning guidance for the number of medical evacuation aircraft needed to support the air assault. To maximize the amount of mission hours they can support the mission, personnel should stage medical evacuation aircraft to support an air assault at the latest possible time. Medical evacuation aircraft should support short distance air assaults from the pickup zone (PZ) or brigade support area. Aircraft may stage at a FARP or use a restricted operations zone to expedite pick up of casualties in long distance air assaults.

9-74. Medical and casualty evacuation aircraft normally are under operational control to the AATF during air assault operations. The air mission commander (AMC) controls the medical evacuation flights to facilitate quick deconfliction of airspace. The AMC clears all medical and casualty evacuation aircraft movements, to include launch and landings. The AATFC may retain launch authority, but the AMC is responsible for medical evacuation.

9-75. Typically, requests for medical or casualty evacuation is over the combat aviation network for the duration of the air assault operation until an evacuation network, if necessary, is established. This ensures good coordination for deconfliction of fires and airspace. When planning medical evacuation operations—
- Send medical evacuation aircraft into secure landing zones (LZs) if possible.
- Integrate attack reconnaissance aviation units to provide escort and LZ overwatch as required.
- Ensure terminal guidance into the LZ.
- Ensure redundant means of communication with the supporting medical evacuation assets throughout the air assault.
- Designate a medical evacuation officer in charge, typically a medical officer from the brigade support medical company, to ride on mission command aircraft to receive and prioritize evacuation mission requests and forward this information to the AMC for launch.

CASUALTY BACKHAUL

9-76. The AATF staff and aviation unit staff plan the combined use of aerial and ground medical and casualty evacuation assets during air assault planning. While assaulting aircraft may backhaul wounded from the LZ, the time required to load and unload casualties could desynchronize the air movement table.

9-77. Casualty evacuation during an air assault may cause delays in air assault missions unless spare aircraft are committed to replace aircraft designated to backhaul casualties. Designating separate casualty evacuation aircraft may prevent delays of follow-on lifts. Procedures for casualty backhaul during an air assault are as follows:
- Medical evacuation request goes to mission command aircraft. The medical officer onboard relays the request to the AMC. If the request is approved, the AMC directs the next serial's last two aircraft (dependent on METT-TC) to move to the LZ casualty collection point to pick up casualties after dropping off personnel.
- All backhauled casualties are taken back to the PZ casualty collection point.
- Backhaul aircraft with casualties notify PZ control they are inbound with casualties.
- Last serial of the final lift makes the final pick up of casualties before the conclusion of the air assault, if necessary.

MEDICAL EVACUATION LANDING ZONE

9-78. During air assault planning, the AATF staff and AMC plan the combined use of air medical evacuation and aerial casualty evacuation landing zones. The AATF plans a means of marking the casualty collection point for air medical or assault aircrew identification. Preferred LZ signaling methods include smoke or panel markers (VS-17 panel marker) during the day and strobe or chemical lights (not blue or green which are not visible under night vision goggles) at night. If air assault crews evacuate casualties, they must know where to take them and how to rejoin remaining lift aircraft for subsequent lifts. Using a backhaul LZ within the normal assault PZ, FARP, or both can minimize disruption of the loading plan while helping maintain serial integrity.

9-79. A dedicated medical evacuation LZ may be used for both air medical and aerial casualty evacuation. Medical evacuation LZ selection and procedures include—

9-80. A dedicated medical evacuation LZ may be used for both air medical and aerial casualty evacuation. Medical evacuation LZ selection and procedures include—
- Select LZs that are level and clear of debris within a 50-meter radius.
- Keep all other light sources away from the LZ unless instructed otherwise by aircrew.
- Once medical evacuation aircraft are inbound, make an estimated time-of-arrival call.

- Ensure personnel communicating with the aircraft at the pickup site have visual on the LZ to confirm the signal or to assist the crew as required.
- Once medical evacuation aircraft has landed, if manned with medical personnel to provide en route care, keep personnel away from the aircraft while the medical personnel come to the patient.
- The unit should provide personnel to assist in loading the patient on the aircraft, and if present, under direction of medical personnel.

Chapter 10
Ground Tactical Plan

The ground tactical plan is the foundation of a successful air assault on which all other air assault planning stages are based. It is the decisive operation for the air assault task force (AATF) because it accomplishes the mission assigned by the higher headquarters. It specifies actions in the objective area that lead to accomplishment of the mission and subsequent operations.

SECTION I – ELEMENTS

10-1. The ground tactical plan may assume a variety of possibilities depending on the commander's evaluation of the mission variables of METT-TC. The ground tactical plan for an air assault contains essentially the same elements as other terrain or enemy-oriented offensive operation. However, the elements of the ground tactical plan are prepared to capitalize on speed and mobility to achieve surprise. The following elements are critical to the planning process.

TASK ORGANIZATION

10-2. *Task organization* is a temporary grouping of forces designed to accomplish a particular mission (ADRP 5-0). When determining the task force organization, air assault planners emphasize the following:
- Maximizing combat power in the assault to heighten the surprise and shock effect, which is especially important if the air assault task force plans to land on or near the objective. Assaulting forces organize on or near the objective are prepared to rapidly eliminate enemy forces, immediately seize objectives, and rapidly consolidate for subsequent operations.
- Ensuring the task force inserts enough force to accomplish initial objectives quickly. To prevent being defeated by repositioning mobile enemy forces, air assault task forces must be massed in the landing zone to build up a significant early combat power capability. If adequate combat power cannot be introduced quickly into the objective area, the air assault force lands away from the objective to build up combat power and then assaults like other combat unit.
- Ensuring the air assault task force commander properly allocates his logistics assets to sustain the task force until follow-on forces arrive.

MISSION STATEMENT

10-3. The mission is the task, together with the purpose, that clearly indicates the action to be taken and the reason therefore. Commanders analyze a mission in terms of the commander's intent two echelons up, along with specified and implied tasks. They consider the mission of adjacent units to understand how they contribute to the decisive operation of their higher headquarters. This analysis produces the unit's mission statement.

10-4. A mission statement is a short sentence or paragraph that describes the organization's essential task (or tasks) and purpose— a clear statement of the action to be taken and the reason for doing so. The mission statement contains the elements of who, what, when, where, and why, but seldom specifies how. It is important to remember that an air assault is a type of operation and not a tactical mission task.

COMMANDER'S INTENT

10-5. The *commander's intent* is a clear and concise expression of the purpose of the operation and the desired military end state that supports mission command, provides focus to the staff, and helps subordinate and supporting commanders act to achieve the commander's desired results without further orders, even

Chapter 10

when the operation does not unfold as planned (JP 3-0). It is critical that the AATF planners receive the commander's intent as soon as possible after the mission is received. Even if the ground tactical plan is not complete, air assault planning often begins after the AATFC issues his intent.

10-6. During planning, the initial commander's intent drives course of action development. In execution, the commander's intent guides disciplined initiative as subordinates make decisions when facing unforeseen opportunities or countering threats.

CONCEPT OF OPERATIONS

10-7. The *concept of operations* is a statement that directs the manner in which subordinate units cooperate to accomplish the mission and establishes the sequence of actions the force will use to achieve the end state (ADRP 5-0). The concept of operations expands on the commander's intent by describing how the commander wants the force to accomplish the mission. It states the principal tasks required, the responsible subordinate units, and how the principal tasks complement one another. Commanders and staff use the operational framework to help conceptualize and describe their concept of operation.

10-8. The operational framework proves the commander with basic conceptual options for visualizing and describing operations in time, space, purpose, and resources. Commanders are not bound by any specific framework for conceptually organizing operations; and use one of three conceptual frameworks listed below or in combination. These operational frameworks apply equally to both operational and tactical actions.
- The deep-close-security framework to describe the operation in time and space.
- The decisive-shaping-sustaining framework to articulate the operation in terms of purpose.
- The main and supporting efforts framework to designate the shifting prioritization of resources.

10-9. The deep-close-security operational framework has historically been associated with terrain orientation but can be applied to temporal and organizational orientations as well. Deep operations involve efforts to prevent uncommitted enemy forces from being committed in a coherent manner. Close operations are operations that are within a subordinate commander's area of operations. Security operations involve efforts to provide an early and accurate warning of enemy operations and to provide time and maneuver space within which to react to the enemy.

10-10. The decisive-shaping-sustaining framework lends itself to a broad conceptual orientation. The *decisive operation* is the operation that directly accomplishes the mission (ADRP 3-0). It determines the outcome of a major operation, battle, or engagement. A *shaping operation* is an operation that establishes conditions for the decisive operation through effects on the enemy, other actors, and the terrain (ADRP 3-0). A *sustaining operation* is an operation at any echelon that enables the decisive operation or shaping operation by generating and maintaining combat power (ADRP 3-0).

10-11. The main and supporting efforts operational framework—simpler than other organizing frameworks—focuses on prioritizing effort among subordinate units. Therefore, leaders can use the main and supporting efforts with either the deep-close-security framework or the decisive-shaping-sustaining framework. The *main effort* is a designated subordinate unit whose mission at a given point in time is most critical to overall mission success. It usually is weighted with the preponderance of combat power (ADRP 3-0). A *supporting effort* is a designated subordinate unit with a mission that supports the success of the main effort (ADRP 3-0). (Refer to ADRP 3-0 for more information.)

DECISIVE-SHAPING-SUSTAINING FRAMEWORK EXAMPLE

10-12. Commanders identify the decisive operation and unit(s) responsible for conducting the decisive operation. This allows them to articulate their shaping operations and the principal task of the units assigned each shaping operation. Commanders complete their CONOPS with sustaining actions essential to the success of decisive and shaping operations.

Decisive Operations

10-13. In figure 10-1, the AATF has been directed by its higher headquarters to conduct an air assault to destroy enemy forces on Objective Horse. The AATFC determines that his decisive operation is the attack to destroy enemy forces on Objective Horse. He further decides that the decisive point of this operation is the successful air assault of his forces into the objective area to destroy the enemy.

Shaping Operations

10-14. In figure 10-1, the AATFC employs his reconnaissance force (to include scouts, CBRN platoon, and UAS) to conduct reconnaissance and surveillance of proposed landing zones and the objective area to identify and target enemy forces near the landing zones and objective vicinities. The field artillery battalion is positioned to provide fires throughout all phases of the operation. It is prepared to deny the enemy's ability to conduct reconnaissance, defeat his strike operations, and neutralize his ability to communicate and command. The supporting aviation unit is prepared to conduct reconnaissance in coordination with the reconnaissance force or provide fires to neutralize or destroy enemy forces in the objective area once they have been identified. The commander considers employing other enablers that may not be in his task force, such as close air support and electronic warfare assets. The purpose of these operations is to set and preserve the conditions on the landing zone and objective area that allow the maneuver forces to launch the air assault and execute a successful attack to destroy the enemy on the objective.

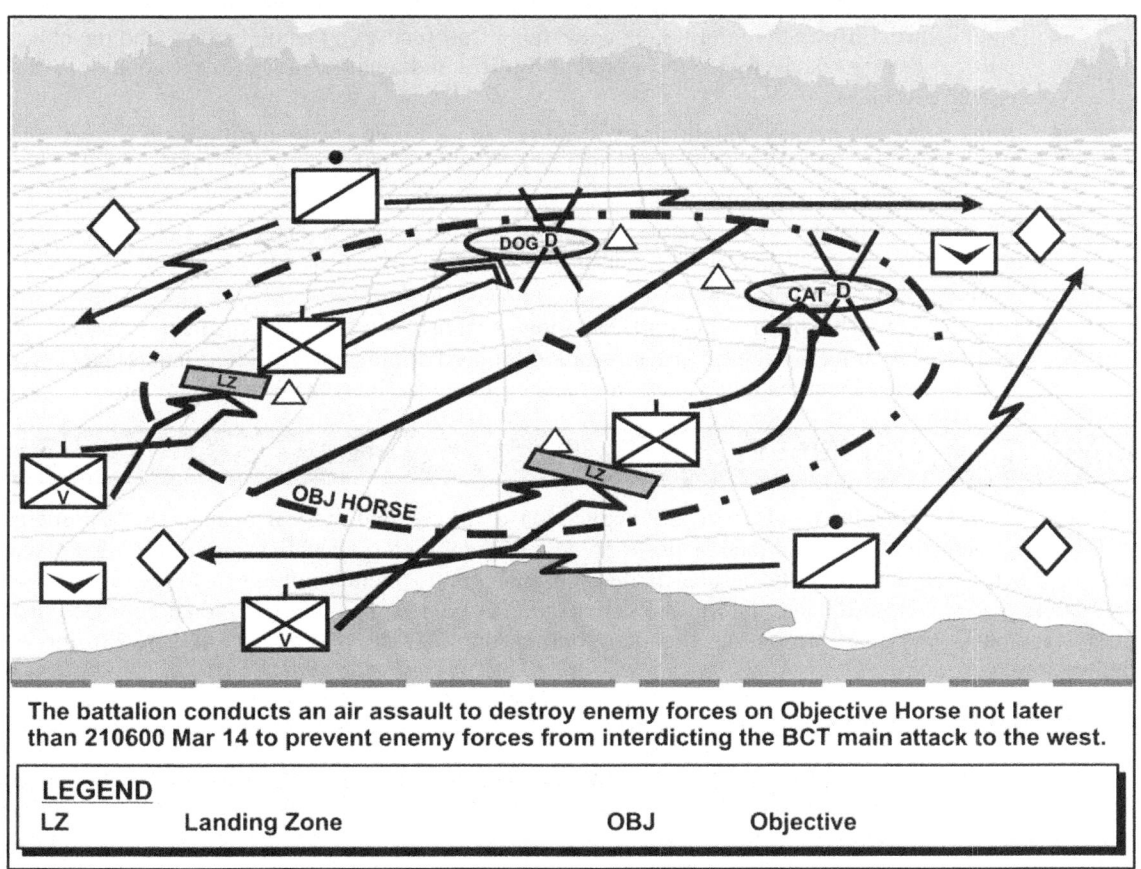

Figure 10-1. Organizational framework

Sustaining Operations

10-15. The AATFC considers how he refuels the supporting aviation unit, resupplies, and provides responsive medical and casualty evacuation to his task force. He determines that this operation may require bringing a forward logistics element from the brigade support battalion and some of its crucial elements forward to conduct casualty evacuation, resupply, and equipment recovery. He positions a forward surgical team or treatment team with a maneuver force to treat casualties before evacuation to a medical treatment facility.

TASKS TO SUBORDINATE UNITS

10-16. The BCT commander through his staff assigns tasks to subordinate units. The assignment of a task includes not only the task (what), but also the unit (who), place (where), time (when), and purpose (why). A task is a clearly defined and measurable activity accomplished by individuals and organizations. Tasks are specific activities that contribute to accomplishing missions or other requirements and direct friendly action. The purpose of each task should nest with completing another task, achieving an objective, or attaining an end state condition.

10-17. Examples of decisive, shaping, and sustaining activities are as follows:
- Mission command. The AATFC deploys in a mission command aircraft to allow the commander to observe and direct the air assault from a forward position.
- Movement and maneuver. Maneuver force conducts an air assault and attack to destroy enemy forces on objectives. Aviation units conduct air insertion of reconnaissance force near the objective area and provide interdiction and close combat attack against identified enemy forces in the objective area.
- Intelligence. Ensures the information collection effort focuses on landing zones and the objective area to identify enemy forces for targeting by fires and aviation assets to set conditions for air assault execution.
- Fires. The field artillery battalion provides fires on identified enemy positions on or near landing zones and the objective to neutralize enemy forces and help set conditions for air assault execution.
- Protection. Suppression of enemy air defenses— plan, synchronize, and execute route lethal suppressive fires and nonlethal suppressive effects on known or suspected enemy air defense positions that are unavoidable.
- Sustainment. The brigade support battalion establishes a forward medical treatment point colocated with the maneuver force. Aviation forward arming and refueling points may colocate with brigade support battalion assets for security and to facilitate the refuel of aircraft.

SECTION II – PLAN DEVELOPMENT

10-18. The AATFC begins to visualize the application of his ground tactical plan to the battlefield by defining the tactical problem and then begins a process of determining feasible solutions with his planning staff using mission analysis. The results of mission analysis (to include intelligence preparation of the battlefield and running estimates) inform the commander as he develops his operational approach that, in turn, facilitates course of action development during the MDMP. (Refer to ADRP 5-0 for more information.)

MISSION ANALYSIS

10-19. Upon receipt of a warning order or mission, the commander supported by his staff, filters relevant information categorized by operational variables into the categories of the mission variables used during mission analysis. The commander uses mission variables to refine his understanding of the situation and to gather relevant information used for mission analysis. Incorporating the analysis of the operational variables with METT-TC ensures the commander considers the best available relevant information about conditions that pertain to the mission. The mission variables of METT-TC consist of mission, enemy, terrain and weather, troops and support available-time available, and civil considerations.

MISSION

10-20. The analysis of the mission is conducted early-on during mission analysis. The mission involves the critical tasks that must be performed. The tasks are either specified tasks stated by the order or implied tasks that the commander must deduce. Mission analysis determines not only what must be accomplished, the intent of the commander ordering the mission (the why of the operation), and the limitations (when, where, how) placed by the higher headquarters, but is the basis for deciding on task organization. Once the mission is analyzed and deductions are made, all other factors are considered in terms of their impact on the mission. It is therefore imperative that the mission be understood.

ENEMY

10-21. Examining enemy factors should be as detailed as possible depending on the time available. General factors to consider are—
- Identification. Size and type of unit (regular or irregular force, or some combination of the two).
- Location. Current and future movement.
- Disposition. Organization or formation.
- Strength. Compared to friendly forces.
- Morale. Esprit, experience, state of training, regular or reserve.
- Capabilities. Electronic warfare, chemical, biological, radiological, and nuclear, air defense, airborne, air movement, attack helicopters, mobility (in comparison to the air assault force).
- Composition. Armored, Infantry, artillery, and sustainment.
- Probable courses of action. Likely mission or objective, probability of achieving it.
- Most dangerous course of action as it applies to the AATF mission and its potential impact on the ground tactical plan.

10-22. When planning an air assault operation, the following factors about the enemy must be considered:
- Air defense weapons and capability.
- Mobility; particularly his ability to react to an air assault insertion.
- Chemical, biological, radiological, and nuclear capability; particularly his ability to influence potential flight routes and landing zones.
- Capability to interdict or interrupt air assault operations with his helicopters or fixed-wing aircraft.

TERRAIN AND WEATHER

10-23. In air assault operations, terrain and weather must be analyzed in terms of their effect on the air assault force. This includes the air assault force's pick up, air movement, insertion, and movement to the final objective, and in terms of the overall influence on aviation operations.

Terrain

10-24. Terrain preparation starts with the situational understanding of terrain through proper terrain analysis. Terrain analysis described in terms of the military aspects of terrain includes observation and fields of fire, avenues of approach, key terrain, obstacles, and cover and concealment (OAKOC).

Observation and Fields of Fire

10-25. Observation is the condition of weather and terrain that permits a force to see friendly, enemy, and neutral personnel, systems, and key aspects of the environment. An assault force's field of fire is directly related to its ability to observe. Considerations related to both enemy and friendly forces and, for air assault operations, include—
- Enemy visual observation or electronic surveillance of pickup zones, flight routes, and landing zones.
- Enhanced friendly observation provided by scout weapons teams and aerial field artillery observation helicopters.
- Ease of navigation along flight routes particularly for night or adverse weather operations.

Avenues of Approach

10-26. Air and ground avenues of approach are considered in both offensive and defensive operations from friendly and enemy viewpoints. A good avenue of approach for air assault forces offers—
- A reasonable degree of mobility and few natural obstacles to the aircraft.
- Little or no canalization.
- Terrain masking that decreases effectiveness of enemy air defense weapons.

Chapter 10

- Cover.
- Concealment.
- Good lines of communication and logistics.
- Ease of linkup with other forces when appropriate.

Key Terrain

10-27. Key terrain is mission-dependent; however, in air assault operations key terrain is not limited to that which influences the ground tactical plan. It must be analyzed in terms of the following actions:
- Pickup zones or landing zones.
- Flight routes.
- Attack weapons team and scout weapons team battle positions.
- Occupation by enemy air defense artillery assets.
- Potential forward arming and resupply points.
- Ground attack positions.

Obstacles

10-28. *Obstacles* are any natural or man-made obstruction designed or employed to disrupt, fix, turn, or block the movement of an opposing force, and to impose additional losses in personnel, time, and equipment on the opposing force (JP 3-15). While most obstacles can be bypassed by air assault forces, obstacles that affect the ground tactical plan must be considered.

Cover and Concealment

10-29. *Cover* is protection from the effects of fires (ADRP 1-02). *Concealment* is the protection from observation or surveillance (ADRP 1-02). Cover and concealment considerations which affect the ground tactical plan include—
- Terrain masking for nap-of-the-earth flight routes and insertions.
- Cover for attack weapons team and scout weapons team positions.
- Landing zones that offer Infantry cover and concealment following insertion.

Weather

10-30. Weather and visibility information is analyzed for trends. (See figure 10-2.) If the operation begins in marginal weather, the commander must consider the possibility that it deteriorates below acceptable limits during the operation. This may result in interrupting helicopter support and requiring changes in planned operations. Considerations include—
- Fog, low clouds, heavy rain, and other factors that limit visibility for aviators.
- Illumination and moon angle during aviation operations with night vision goggles.
- Ice and sleet, and freezing rain that degrades aerodynamic efficiency, and impact the ground tactical plan.
- High temperatures or density altitudes that degrade aircraft engine performance and lift capability.
- Darkness, normally an advantage to well-trained aviators and Soldiers.
- High winds (large-gust spreads).
- Weather conditions that create hazards on pickup zones and landing zones, such as blowing dust, sand, or snow.

Ground Tactical Plan

FORMAT	MAXIMUM WINDS
Weather forecasts are received in the following format: Ceiling. Visibility. Weather (for example, clear, fog, rain, snow). Additional information as requested by the S-2 intelligence staff officer.	Observation helicopter (OH): 30 knots. Utility helicopter (UH):40 knots. Cargo helicopter (CH):60 knots. ***Note**. Gusting winds, in excess of 15 knots over the lull wind, may avoid UH usage. Significant weather patterns (which limit operations) are moderate turbulence and icing.
CONSIDERATIONS	
Allowable Weather Limits (Applicable to combat operations and tactical training at a military airfield). Visibility: 1/2 mile. Ceilings: Clear of clouds.	Extremes Limiting Tactical Air Ceiling:1,000 feet.* Visibility: 2 miles.* ***Note**. Operational design of a A-10, close air support aircraft. Other type aircraft require better weather conditions.

Figure 10-2. Weather data.

TROOPS AND SUPPORT AVAILABLE

10-31. Troops and support available include the number, type, capabilities, and condition of assault and support troops and support aviation available to conduct the air assault operation. Critical considerations to the air assault operation include:

- The AATF should have enough combat power to seize initial objectives and protect the landing zones until follow-on echelons arrive in the objective area.
- Assault (lift) helicopter capability is the single most important variable in determining how much combat power can be introduced into the objective area.

10-32. Aircrew endurance must be considered. The aviation task force standard operating procedures (SOP) outlines aircrew duty day and flying hour limits. If those limits are exceeded during a single period, then degraded aircrew performance or limited aircrew availability can be expected on the following days.

TIME AVAILABLE

10-33. The commander assesses the time available for planning, preparing, and executing tasks and operations. This includes the time required for pick up, air movement, insertion, movement to the final objective, and delivery of follow-on forces in relationship to the enemy and conditions. Critical considerations to the air assault operation include:

- Air assault planning must be centralized and precise, and takes more time than that for other operations. Time must be made available for air-ground operations preparation, planning, and rehearsals.
- Allot additional planning time for night operations and those involving multiple pickup zones or multiple landing zones.
- The AATFC must allow adequate time to ensure that all subordinates units, particularly aviation aircrews, are thoroughly briefed. Viable SOPs and previous training significantly reduce briefing time.

CIVIL CONSIDERATIONS

10-34. The ability to analyze civil considerations to determine their impact on operations enhances several aspects of the air assault operation to include air movement, insertion into the objective area, movement to the final objective, and follow-on operations. (Refer to ATP 2-01.3 for more information.) Civil considerations comprise six characteristics, expressed in the memory aid ASCOPE—

Chapter 10

- Areas.
- Structures.
- Capabilities.
- Organizations.
- People.
- Events.

ASSAULT OBJECTIVE AND LANDING ZONE DEVELOPMENT

10-35. The ground tactical plan for an air assault operation contains essentially the same elements as other attacks but differs in that it is prepared to capitalize on speed and mobility to achieve surprise. Assault echelons are placed on or near the objective and organized to be capable of immediate seizure of objectives and rapid consolidation for subsequent operations. If adequate combat power cannot be introduced quickly into the objective area, then the air assault force must land away from the objective and build up combat power. The air assault force then assaults like other Infantry units and the effectiveness of the air assault operation is diminished. (Refer to FM 3-21.20, FM 3-21.10, and FM 3-21.8 for more information.)

10-36. The scheme of maneuver may assume a variety of possibilities depending on the commander's evaluation of METT-TC to include, in particular, the availability of landing zones in the area. The plan includes—
- Missions of all task force elements and methods for employment.
- Areas of operations with graphic control measures.
- Task organization to include command relationships.
- Location and size of reserves.
- Fire support to include graphic control measures.
- Sustainment.

SECTION III – AIR-GROUND OPERATIONS

10-37. Employing aviation forces with ground maneuver forces requires detailed integration and coordinated force-oriented control measures to support ground maneuver while minimizing fratricide risks. This section discusses the integration of air and ground maneuver, close combat attack, close air support, and unmanned aircraft system employment during air-ground operations. (Refer to FM 3-04.111 for more information.)

EFFECTIVE INTEGRATION

10-38. Integrating effective air and ground maneuver forces begins at the AATF and continues down to the lowest unit level. In an air assault, the AATF plans and coordinates with the supporting combat aviation brigade through their aviation liaison officer and the ADAM/BAE to support the ground tactical plan. Integration should start at the home station with implementation of effective SOPs, habitual relationships, and training if possible.

10-39. Integration involves merging the air and ground fights into one to apply proper aviation capabilities according to the supported AATFC's intent. Integration ideally begins early in the planning process with the involvement of the ADAM/BAE. The ADAM/BAE advises the AATFC on aviation capabilities and the best way to use aviation to support mission objectives. Ensuring the aviation liaison officer or brigade aviation element passes along the task and purpose for aviation support and continually provides updates as needed is of equal importance. Simply stated, ensuring the aviation brigade and subordinate unit staffs fully understand the AATF scheme of maneuver and commander's intent is critical to successful air-ground operations.

10-40. Employing attack reconnaissance aviation with ground maneuver forces requires coordinated force-oriented control measures and the CCA 5-Line attack brief allowing aviation forces to support ground maneuver with direct fires while minimizing fratricide risks. Aviation liaison officers should identify early in the planning process the minimum AATF graphics required for operations such as boundaries, phase lines, attack by fire positions, and objectives. Brigade aviation element and liaison officer personnel should ensure that supported units are familiar with close combat attack request procedures and marking methods.

CLOSE COMBAT ATTACKS

10-41. A close combat attack is a coordinated attack by Army attack reconnaissance aviation aircraft (manned and unmanned) against targets that are in close proximity to friendly forces. In most instances, the attack aviation may already occupy holding areas, battle or support by fire positions or are in overwatch of the ground maneuver force as it begins its assault. The AATF employs close combat attack procedures to ensure that these aviation fires destroy the enemy with minimal risk to friendly forces.

10-42. Close combat attack is not synonymous with close air support flown by joint aircraft. Due to capabilities of the aircraft and the enhanced situational awareness of the Army aircrews, terminal attack control from ground units or controllers is not required. The most important factor of successful close combat attacks is positive and direct communication between aviation and ground elements. Aviators and ground elements need to understand the following to employ a successful close combat attack. (Refer to FM 3-04.126 for more information.)

CLOSE COMBAT ATTACK REQUEST

10-43. When providing support to ground maneuver elements, Army aviation will operate on that echelon's command network unless directed otherwise. The CCA 5-Line attack brief (Format 22) is the standard brief for Army rotary wing and organic-armed UASs. The format is similar to the 5-Line RW CAS attack brief, but will be prefaced with "Fire mission" vice a type of control. At check-in, Army attack and scout weapons teams will brief the Format 23. CCA Check-in, Aircraft Transmits to Ground Unit. (Refer to ATP 3-09.32 for more information.)

10-44. Any element in contact uses the CCA 5-Line attack brief to initiate the close combat attack. The CCA 5-Line attack brief allows the ground maneuver forces to communicate and reconfirm to the aircraft the exact location of friendly and enemy forces. The procedure remains the same regardless of the type of unit in contact or the responding aviation element. The ground commander owning the terrain clears fires during the close combat attack by giving aircrews the situational awareness of the location of friendly elements. The ground commander deconflicts the airspace between indirect fires, close air support, UAS and the close combat attack aircraft.

10-45. Transmission of the brief constitutes clearance to fire except in a danger close situation. For danger close fire, the ground commander on the scene accepts responsibility for increased risk. Danger close must be declared in the Line 5 when applicable by stating "Cleared Danger Close" and passing the initials of the ground commander on scene.

10-46. After receiving the request for close combat attack, the aircrew informs the ground maneuver force leader of the battle position, attack- or support by fire position (or series of positions) the team is occupying, and the location from which the attack aircraft engages the enemy with direct fire. The size of this position varies depending on the number of aircraft using the position, the size of the engagement area, and the type of terrain.

10-47. The position must be close enough to the requesting unit to facilitate efficient target handover. Aircraft leaders normally offset the position from the flank of the friendly ground position. This helps to ensure that rotor wash, ammunition casing expenditure, and the general signature of the aircraft do not interfere with operations on the ground. The offset position allows the aircraft to engage the enemy on its flanks rather than its front. It reduces the risk of fratricide along the helicopter gun-target line. (Refer to ATP 3-09.32 for more information.)

GROUND MANEUVER FORCE AND TARGET MARKING

10-48. Marking methods for identifying targets and friendly positions vary from one ground maneuver force to another. The close combat attack request should include a detailed description of all friendly locations and target locations in relation to friendly positions. It should include the target description and how it is marked.

Chapter 10

10-49. For mutual protection and clarity on the appropriate target, the ground maneuver force does not mark the target until requested by the aviation element. This in no way restricts the ground maneuver force from returning fire from the enemy. However, the ground maneuver force should consider that the aircrews may not be able to distinguish the correct target from other fires if they mark the target with fire. Ground maneuver forces should have multiple means of marking their positions. If the target is marked by fire, the aviation element requests the ground maneuver force to stop marking. The aviation element calls when clear of the area and reports estimated battle damage assessment.

10-50. The close combat attack cannot be conducted without positive identification of friendly and enemy forces by both the ground and aviation commander before attack aviation aircraft opens fire. The aviation element tailors its attack angles and weapon selections based upon the target and friendly unit proximity to the target.

TARGET HANDOVER

10-51. The rapid and accurate marking of a target is essential to a positive target handover. Aircraft conducting close combat attacks normally rely on a high rate of speed and low altitude for survivability in the target area. As such, the aircrew generally has an extremely limited amount of time to acquire both the friendly and enemy marks. It is essential that the ground maneuver force has the marking ready and turned on when requested by the aircrew.

10-52. Attack reconnaissance aircrews use both thermal sights and NVGs to fly with and acquire targets. After initially engaging the target, the aircrew generally approaches from a different angle for survivability reasons if another attack is required. The observer makes adjustments using the eight cardinal directions and distance (meters) in relation to the last round's impact and the actual target. At the conclusion of the close combat attack, the aircrew provides its best estimate of battle damage assessment to the unit in contact.

BATTLE DAMAGE ASSESSMENT AND REATTACK

10-53. After the attack aircraft complete the requested close combat attack mission, the aircrew provides a battle damage assessment to the ground commander. Based on his intent, the ground maneuver commander determines if another attack is required to achieve his desired end state. The close combat attack operation can continue until the aircraft have expended all available munitions or fuel. However, if the air mission commander receives a request for another attack, he must carefully evaluate his ability to extend the operation. If not able, he calls for relief on station by another attack team if available. It is unlikely that the original team has enough time to refuel, rearm, and return to station.

CLEARANCE OF FIRES

10-54. During an air assault with numerous aircraft in the vicinity of the landing zone, it is critical that procedures are in place to deconflict airspace between aircraft and indirect fires, considerations include—
- Ensure aircrews have the current and planned indirect fire positions (to include mortars) supporting the air assault before the mission.
- Plan for informal airspace coordination areas and check firing procedures and communications to ensure artillery and mortars firing from within the landing zone do not endanger subsequent serials landing or departing, close combat attack, or close air support.
- Ensure at least one of the aviation team members monitors the fire support net for situational awareness.
- Advise the aviation element if the location of indirect fire units changes from that planned.
- Ensure all participating units are briefed daily on current airspace control order or air tasking order changes and updates that may affect air mission planning and execution.
- Ensure all units update firing unit locations, firing point origins, and final protective fire lines as they change for inclusion in current airspace control order.

10-55. The AATFC or ground commander can establish an informal airspace coordination area. For example, he can designate that all indirect fires be south of and all aviation stay north of a specified gridline for a specific period. This is one method for deconflicting airspace while allowing both indirect fires and attack aviation to attack the same target. The ground commander then can cancel the informal airspace coordination area when the situation permits. (Refer to FM 3-52 for more information.)

CLOSE AIR SUPPORT

10-56. Close air support is air action by fixed- and rotary-wing aircraft against hostile targets that are in close proximity to friendly forces and that require detailed integration of each air mission with the fire and movement of those forces. Like close combat attack, close air support can be conducted at any place and time friendly forces are in close proximity to enemy forces based on availability. All leaders in the AATF should understand how to employ close air support to destroy, disrupt, suppress, fix, harass, neutralize, or delay enemy forces. Nomination of close air support targets is the responsibility of the commander, air liaison officer, and S-3 at each level. The AATF may receive close air support from USAF, USN, USMC, or multinational force. (Refer to JP 3-09.3 for more information.)

CAPABILITIES AND EMPLOYMENT

10-57. In some cases, USAF aircraft are available to provide close air support. Requests for these aircraft are processed through the TACP colocated with the BCT main command post. The TACP is organized as an air execution cell capable of requesting and executing Type 2 or 3 terminal attack control of close air support missions. The manning of the cell depends on the situation but, at a minimum, includes an air liaison officer and a JTAC. To make a recommendation to the commander regarding the use of close air support aircraft, the leader on the ground should be familiar with the characteristics of the aircraft predominantly used in the close air support role. (Refer to ATP 3-09.32 for more information.)

BRIEFING FORMAT

10-58. Two types of close air support requests are listed as follows:
- Preplanned requests that may be filled with either scheduled or on-call air missions. Those close air support requirements foreseen early enough to be included in the first air tasking order distribution are submitted as preplanned air support requests for close air support. Only those air support requests submitted in sufficient time to be included in the joint air tasking cycle planning phases and supported on the air tasking order are considered preplanned requests.
- Immediate requests that mostly are filled by diverting preplanned missions or with on-call missions. Immediate requests arise from situations that develop outside the air tasking order planning cycle.

10-59. The air liaison officer and JTAC personnel in the TACP are the primary means for requesting and controlling close air support. However, forces may have joint fires observer certified personnel who can request, adjust, and control surface-to-surface fires, provide targeting information in support of Type 2 and 3 close air support terminal attack controls, and perform autonomous terminal guidance operations. (Refer to ATP 3-09.32 for more information.)

UNMANNED AIRCRAFT SYSTEMS

10-60. Unmanned aircraft system (UAS) operations provide surveillance capabilities to enhance the AATFC's situational awareness as he plans, coordinates, and executes the air assault. The commander can employ UAS from his organic elements or he can request to have direct access to real-time feeds from additional UAS support from his higher headquarters. They are particularly effective when employed together with ground and attack reconnaissance elements as a team during shaping operations in which the commander is trying to create the conditions for successful air assault execution. (Refer to chapter 4 of this publication for UAS discussion.)

Chapter 10

SECTION IV – EXECUTION

10-61. An AATF is normally a highly tailored force specifically designed to hit hard and fast and is employed in situations that provide the task force a calculated advantage due to surprise, terrain, threat, or mobility. The following employment considerations govern the execution of the air assault operations.

CONDUCT OF THE AIR ASSAULT

10-62. The AATF normally should be assigned only missions that take advantage of its superior mobility and should not be employed in roles requiring deliberate operations over an extended period. The basic principles that apply to the conduct of air assault operations include—

- Air assault forces always fight as a combined arms team.
- Availability of critical aviation assets is a major factor in an operation.
- Air assault planning must be centralized and precise; execution must be aggressive and decentralized.
- Air assault operations may be conducted at night or during adverse weather, but require more planning and preparation time in those cases.
- Assault force tactical integrity must be maintained throughout an air assault. Squads are loaded intact on the same helicopter, with platoons located in the same serial ensures fighting unit integrity upon landing.
- Fires must provide for suppressive fires along flight routes and near landing zones. Priority for fires must be to the suppression of enemy air defenses.
- Infantry operations are not fundamentally changed by integrating aviation units with Infantry; tempo and distance are changed dramatically.
- An air assault operation may be the decisive operation; examples include river crossings, seizure of key terrain, raids, and security area actions.
- An AATF is employed most effectively in environments where limited lines of communication are available to the enemy, where he lacks air superiority and effective air defense systems.

BUILDUP OF COMBAT POWER

10-63. The availability of aviation assets is normally the major factor in determining AATF task organization. The AATF must provide a mission-specific balance of mobility and combat power to include sustainment of combat power. Considerations for rapid massing and shifting of combat power during the conduct of an air assault include—

- Delivery to the objective area as soon as possible, consistent with aircraft and pickup zone capacities, to provide surprise and shock effect.
- Arrival intact at the landing zone with assault force tailored to provide en route security and protection from the pickup zone, throughout the entire flight route, and at the landing zone.
- Nonstandard command relationship, attached for movement, used extensively during air assault operations. Elements, to include field artillery, air defense artillery, intelligence, and engineer may be attached to maneuver elements for movement only, to facilitate mission command, maneuver, and security.
- Task-organized with sufficient combat power to seize initial objectives and protect landing zones, and with sufficient sustainment for rapid tempo until follow-on or linkup forces arrive, or until the mission is completed.

Chapter 11
Landing Plan

The landing plan supports the ground tactical plan. It provides a sequence for arrival of units into the area of operation, ensuring that all assigned units arrive at designated locations and times prepared to execute the ground tactical plan. General considerations to develop the landing plan follow.

SECTION I – LANDING ZONE SELECTION

11-1. Landing zones usually are selected by the AATFC or his S-3 based on technical advice from the air mission commander or the aviation liaison officer. The section addresses keys considerations for landing zone selection.

CRITERIA FOR SELECTING LANDING ZONES

11-2. Air assault landing zone selection is based on user requirements; type of environment; availability; adequacy; security of landing zones on or near the objective area, or away from the objective area; threats to the objective area; and aircraft/aircrew capability. Criteria for selecting landing zones include—

- Location. In general, two options are viable when selecting landing zones—land on the objective or land away from the objective. The selection of either option is METT-TC dependent.
- Capacity. Size determines how much combat power can be inserted at one time and the need for additional landing zones or time separation between serials.
- Types of loads. External loads generally require larger landing zones than landing zones for personnel alone.
- Elevation. The altitude of potential landing zones may not be supportable due to operating restrictions of certain aircraft.
- Alternates. An alternate landing zone should be planned for each primary landing zone to ensure flexibility to support the mission.
- Enemy composition, disposition, and capabilities. Landing zone considerations include enemy force concentrations, weapons systems, and their capability to react to an air assault task force landing nearby.
- Cover and concealment. Select landing zones to deny enemy observation and acquisition of friendly ground and air elements while they are en route to, from, or in the landing zone.
- Obstacles. If possible, the air assault task force should land on the enemy side of obstacles when attacking to negate their effectiveness. The air assault task force should consider using obstacles to protect landing zones from the enemy at other times.
- Landing point. Landing zones should be generally free of obstacles. The landing point or touchdown point must be free of obstacles (any object or hole greater than 18 inches high or deep). (Refer to FM 3-21.38 for more information.)
- Identification from the air. If possible, landing zones should be easily identifiable from the air or marked by friendly reconnaissance forces that have reconnoitered the landing zone.
- Approach and departure routes. If possible, approach and departure air routes should avoid continued exposure of aircraft to enemy fire.
- Weather. Reduced visibility or strong winds may cause aircraft to avoid or limit the use of primary or alternate landing zones.

LOCATION OF LANDING ZONES

11-3. The AATF plans to land on or near the objective when the assault force is able maximize combat power in the landing zone to rapidly eliminate enemy forces, immediately seize objectives, and rapidly consolidate for subsequent operations. Considerations for landing on or near the objective include—
- Air assault task force is assigned a terrain-oriented mission.
- Commander has accurate up-to-date intelligence on the enemy.
- Commander has accurate intelligence on terrain (especially landing zones), weather is favorable, and suitable landing zones are available on or near the objective.
- Shaping operations have set conditions for air assault execution.
- Time in which to accomplish the overall mission is limited.
- Civilian population is known to be supportive of U.S. presence in the area of operation.

11-4. When adequate combat power cannot be introduced quickly into the objective area, the air assault force lands away from the objective to build up combat power and then assaults to seize objectives. Considerations for landing away from the objective include—
- Air assault task force is assigned an enemy-oriented mission.
- Commander has incomplete or unknown intelligence on the enemy.
- Commander has incomplete information on terrain (especially landing zones), weather is not favorable, or no suitable landing zones are available near the objective.
- Shaping operations have not set conditions for air assault execution or conditions cannot be verified.
- Time is available upon landing in the landing zone to develop the situation.
- Civilian population is unknown or hostile to U.S. presence in the area of operation.

NUMBER OF LANDING ZONES

11-5. The AATFC decides whether to use a single landing zone or multiple landing zones. A large number of landing zones for an air assault increases the tactical risk and complexity of the operation as well as the difficulty of setting conditions at each landing zone before landing. The AATFC should plan for one primary landing zone and one alternate landing zone according to the maneuver force regardless of proximity to the objective. He should plan for more than one primary and one alternate landing zone according to maneuver force only after careful analysis of the mission variables to determine if sufficient forces are available to conduct shaping operations at each landing zone.

11-6. Using a single landing zone—
- Requires less planning and rehearsal time.
- Allows concentration of combat power in one location.
- Facilitates control of the operation.
- Concentrates supporting fires in and around the landing zone.
- Requires fewer attack helicopters for security.
- Provides better security for subsequent lifts.
- Reduces the number of air routes in the objective area, making it more difficult for the enemy to detect the air assault operation.
- Centralizes required resupply operations.

11-7. Using multiple landing zones—
- Avoids grouping units in one location, which creates a lucrative target for enemy mortars, artillery, and close air support.
- Allows rapid dispersal of ground elements to accomplish tasks in separate areas.
- Reduces the enemy's ability to detect and react to the initial lift.
- Forces the enemy to fight in more than one direction.
- Reduces the possibility of troop congestion in one landing zone.

Landing Plan

- Eliminates aircraft congestion on one landing zone.
- Makes it difficult for the enemy to determine the size of the air assault force and the exact location of supporting weapons.

SECTION II – LANDING ZONE UPDATES

11-8. Just before the start of the air movement and just before the lift aircraft reach the release point, the attack reconnaissance aircraft or unmanned aircraft system (UAS) provide a landing zone update to the AATF, informing the AATFC ground commander, and air mission commander of the status of enemy activity on the landing zone. The requirement for a landing zone update is METT-TC dependent and based on the need to preserve surprise on the objective. The manner in which the landing zone update is conducted should not divulge the exact location of the landing zone.

LANDING ZONE CONDITION

11-9. The landing zone is considered cold if no enemy activity is observed. If the landing zone is cold, the air assault is executed as planned. The landing zone is considered hot if enemy activity is occurring on or near the landing zone. If the landing zone is hot, the attack reconnaissance aircraft provide a situation report consisting of enemy activity, their actions toward the enemy and a recommendation for using the alternate landing zone. Based on the recommendation of the attack reconnaissance aviation element, the AATFC decides whether to use an alternate landing zone. As part of the mission analysis and rehearsal process, aircrews rehearse and execute the air movement using an alternate landing zone.

FIXED-WING SUPPORT

11-10. When available, fixed-wing aircraft can be used to provide a landing zone update or to eliminate enemy activity. As long as lift aircraft or attack reconnaissance aircraft possess the proper communication capabilities, fixed-wing aircraft can relay the update directly to the AATFC. If these capabilities are not present, fixed-wing aircraft may relay the update to a command post that then relays the update to the AATFC. The plan must account for time needed to relay the update to all parties.

UNMANNED AIRCRAFT SYSTEM SUPPORT

11-11. UAS can be employed to monitor and relay the updated status of the landing zone and surrounding area during the air movement phase. This early information gives the AATFC more time to adjust plans if required. UAS that fly at higher altitudes may observe with negligible risk of revealing landing zone or objective locations.

SECTION III – HOT LANDING ZONE CONSIDERATIONS

11-12. Sometimes the presence of enemy activity is unknown or unclear until the first aircraft lands in the landing zone. A unit should develop and rehearse its plan for reacting to enemy contact in that situation.

SCENARIOS

11-13. The enemy may employ one or a combination of the following actions to oppose landing operations:
- Conduct a near ambush.
- Conduct a far ambush.
- Deliver indirect fires by mortars, artillery, or rockets directed by an observer that can see the landing zone.
- Emplace obstacles, such as antipersonnel mines, booby traps, or other barriers.

11-14. The AATFC considers five options in response to a hot landing zone. They are—
- Fight through the contact.

Chapter 11

- Divert to the alternate landing zone.
- Abort remaining serials.
- Slow airspeeds to delay serials.
- Racetrack serials.

11-15. Racetracking is considered a high-risk option. All serials orbit at their current position. Once the enemy has been neutralized or destroyed on the landing zone, the air assault resumes in the order outlined in the air movement table. The air mission commander determines whether enough fuel, spacing, and time is available between serials to conduct this option and advises the AATFC accordingly.

11-16. The AATFC makes the final decision on all options involving a hot landing zone, and the air mission commander and ground commander execute. Whether landing away from or on the objective, it is important that primary and alternate landing zones are mutually supporting to allow the AATFC to shift the main effort if needed.

REACTION TO ENEMY CONTACT AWAY FROM THE OBJECTIVE

11-17. When landing away from the objective, ground maneuver forces can more readily divert to an alternate landing zone. In doing so, the main effort of the decisive operation may be shifted to the force landing at the alternate landing zone, and the force at the hot landing zone may be extracted or continue to fight through the enemy contact. If the alternate landing zone is hot as well, the AATFC should choose which force to designate as the main effort to accomplish the mission.

11-18. A force that encounters a near ambush, unless extremely successful in counteracting that ambush, is usually extracted, reorganized, and reinserted into an alternate landing zone to continue the mission. A force that encounters a far ambush, hostile indirect fires, or obstacles usually continues its mission.

REACTION TO ENEMY CONTACT ON THE OBJECTIVE

11-19. When landing on the objective, units react to contact and fight through. Because the landing zone is on the objective, fighting for control of the hot landing zone is critical to mission accomplishment and continuing the assault is the priority. The unit on the hot landing zone may be directed to fix the enemy, while the main effort is shifted to the unit that lands at an alternate landing zone and fights through to the objective.

11-20. If the alternate landing zone is hot as well, the AATFC should choose which unit to designate as the main effort to accomplish the mission. Given the overall mission, breaking contact or extraction is not likely for units caught on a hot landing zone. In cases other than a near ambush, units fight through enemy contact and continue the mission without diverting serials to an alternate landing zone.

SECTION IV – PREPARATION AND SUPPORTING FIRES

11-21. Preparation fires are planned for each landing zone so they can be executed if needed. However, it is desirable to make the initial assault without preparation fires to achieve tactical surprise. Planned fires for air assault operations should be intense and short but with a high volume of fire to maximize the surprise and shock effect.

11-22. Ground forces land ready to fight, with the integrated support of close air support, close combat attack, and indirect fires. Supporting fires, direct or indirect, are directed and cleared on the landing zone by the ground commander.

11-23. Factors to consider when developing fire support plans are listed as follows:
- Deception. False preparations can be fired into areas other than the objective or landing zone area to deceive enemy forces if rules of engagement allows. For example, some rules of engagement might prevent any unobserved fires.
- Duration of preparation fires. A preparation of long duration may reduce the possibility of surprise. The preparation fires should begin as the first aircraft of the first lift crosses the release point and end just before the first aircraft lands.

- Availability of fire support assets. The ground commander coordinates with the artillery unit to arrange the preparation of units that can fire. In some cases, where an air assault is executed across extended distances, preparation fires by close air support or attack helicopters may be the only viable alternative.
- Objective area fires. A known or suspected enemy force in the landing area, regardless of size, warrants preparation fires.
- Effects of ordnance on the landing zone. Some ordnances used in preparation fires (such as artillery, bombs, or infrared illumination) may be undesirable since they can cause craters, downed trees, fires, and landing zone obscuration.
- Scheduling fires. Fires are scheduled to be lifted or shifted to coincide with the arrival times of aircraft formations.
- Collateral damage. The unintentional or incidental injury or damage to persons or objects that would not be lawful military targets in the circumstances ruling at the time. Such damage is lawful so long as it is not excessive in light of the overall military advantage anticipated from the attack. (Refer to JP 3-60 for more information.)
- Positive control measures. Control measures must be established for lifting or shifting fires.
- Additional considerations include:
  - The ground commander clears all ground, air, and indirect fires inside the airhead line.
  - Door gunners in assault aircraft fire only at the base of the tree line to avoid fratricide of overwatching gunships.
  - As long as the air assault continues, attack aviation works for the AATFC. Attack reconnaissance aviation is placed in direct support of the ground commander for air assault security and close combat attack within the objective area. The ground commander passes control of attack aviation aircraft to subordinate commanders for close combat attack. Once the threat is eliminated, attack aircraft are passed back to the ground commander's control. Only a ground commander can clear Apache or Kiowa fires into the tree line assaulted by friendly forces. (Refer to chapter 6 of this publication for more information.)
  - Indirect fires on the tree line being assaulted by friendly forces are always treated as danger close. (In other words, in a right door exit, a fire mission into the right tree line would be danger close.)
  - Know the locations of all friendly forces in the area, to include reconnaissance and long-range surveillance units, pathfinders, and special operations forces.

SECTION V – LANDING SITE OPERATIONS

11-24. Aircraft formations on the landing zone should facilitate a rapid exit from the aircraft, an orderly departure off the landing zone, and an organized deployment for the assault. The number and type of aircraft and the configuration and size of the landing zone may dictate the formation. (See chapter 12 for a discussion of standard flight and landing formations.) If contact is expected in the landing zone, elements must land ready to fight and maneuver in all directions.

LANDING ZONE AND OBSTACLE MARKINGS

11-25. For daylight operations, pathfinders use panels or some other minimal identification means to mark landing zones. Smoke might also be used to identify a landing zone and assist the pilot in determining wind conditions. However, smoke is also easily identified by the enemy. For daylight operations, mark the number one landing point using a single VS-17 panel, with the international orange side visible. Other touchdown points might be marked, as coordinated. Mark obstacles using the cerise colored side of the panel. For night operations, use chem-lights, lanterns, field expedients, or other methods to show the direction of landing and to mark individual landing points. For day and night air assault operations, mark all obstacles. (Refer to FM 3-21.38 for more information on daylight landing formations.)

11-26. At night, pathfinders can use lights of different colors (except red, which marks obstacles) to designate different helicopter sites or to separate flights within a larger formation. A lighted "T" or inverted

Chapter 11

"Y" indicates both the landing point for the lead helicopter of each flight and the direction of approach. Other lights mark touchdown points for the other helicopters in the flight. Each helicopter should land with its right landing gear or its right skid 5 meters left of the lights. Large cargo helicopters (CH-47) land 10 meters to the left of the lights. (Refer to FM 3-21.38 for more information on night landing formations.)

11-27. For security, pathfinders and the ground unit turn off, cover, or turn all lights upside down until the last practical moment before a helicopter arrives. Then they orient the lights in the direction from which the lead helicopter is approaching, and a signalman directs its landing.

Note: Because the marking lights could be too bright for the aircrew member's night vision goggles, crew members might have to look under the goggles to distinguish the colors. Also, aircrew members wear night vision goggles with filtered lenses. These filters do not allow the aircrews to see blue or green chem-lights. Colors such as yellow, orange, red, and infrared can be seen by pilots wearing ANVIS.

11-28. During daylight landing operations, pathfinders use red-colored panels or other red, easily-identifiable means to mark any hard-to-detect, impossible-to-remove obstacles such as wires, holes, stumps, and rocks. During nighttime, pathfinders use red lights to mark any obstacles within the landing site that they cannot reduce or remove.

11-29. In most combat situations, the need for security keeps pathfinders from using red lights to mark treetops on the departure end of a landing zone. However, in training or in a rear area landing site, they do use red lights. If they cannot mark obstacles or hazards, they must fully advise aviators of existing conditions by ground to air radio. In any case, the pathfinder landing site leader makes sure that pathfinders mark the most dangerous obstacles first and, if possible, that they remove them.

11-30. Pathfinders have a limited capability to secure a landing site. If they precede the initial assault elements into a landing site, Soldiers from the supported ground unit can go with them for security. If required to do so by the supported unit, pathfinders can mark initial assembly points for soldiers, equipment, and supplies. They should choose locations that help ensure the quick, efficient assembly and clearing of the helicopter landing site. If the unit uses assembly areas, the ground unit commander selects their locations. If needed, supported ground unit Soldiers go with the pathfinders to reconnoiter and mark the unit assembly areas, set up assembly aids, act as guides, and help with landing and unloading operations. Having this help ensures that the pathfinders can rapidly clear soldiers, supplies, and equipment from the landing points.

11-31. A landing zone formation may not have standardized distances between aircraft due to the size or terrain on the landing zone. The goal in landing aircraft successfully is to select a safe landing area as close to cover and concealment as possible to reduce Soldier exposure. If possible, the aircraft formation on the pickup zone is the same as the landing zone. This provides Soldiers and leaders a preview of the landing zone landing formation and gives them an idea of their location upon landing in relation to other elements.

11-32. The lead elements lifted into the landing zone are responsible for clearing the landing zone to support follow-on lifts. This can be accomplished using a number of methods, which are entirely METT-TC dependent. The most common method for clearing the landing zone is to assign assault objectives, which requires subordinate units to move through an assigned area to clear enemy forces before reaching their final objective.

EXITING THE AIRCRAFT

11-33. The two methods for exiting a UH-60 aircraft are the one-side off-load and the two-side off-load. Soldiers exiting a CH-47 do so from the rear ramp. In each method, Soldiers must be careful to avoid the main and tail rotors of the aircraft they are exiting and the rotors of other aircraft in their serial. The separation between serials and the number of serials that can fit into the landing zone at one time are critical planning considerations when determining the aircraft exiting method.

11-34. As part of an air assault, the mission may require the application of the fast-rope insertion and extraction system (FRIES) for small units to infiltrate or insert into a confined area where a helicopter is unable to land as an alternative method for exiting an aircraft. FRIES is the fastest method of deploying

Landing Plan

Soldiers from a rotary-wing aircraft that are unable to land. FRIES is not approved for Army-wide use and is restricted to special operations forces and long-range surveillance units. (Refer to FM 3-05.210 for more information.)

ONE-SIDE OFF-LOAD

11-35. In this method, Soldiers exit from either the right or left side of the aircraft. (See figure 11-1.) Soldiers exiting the aircraft should step outward and take up a prone position, forming 180-degree security on that side of the aircraft yet remaining under the main rotor system and outside the landing gear of the aircraft. Soldiers should remain in the prone position until the aircraft lifts off before departing the landing zone. The chalk leader directs his chalk to move to the nearest covered and concealed position according to the landing plan or SOPs.

11-36. A unit plans to execute a one-side off-load on the side away from known or potential enemy positions but may be forced to exit the aircraft on the opposite side due to the enemy or other METT-TC considerations once the aircraft has landed.

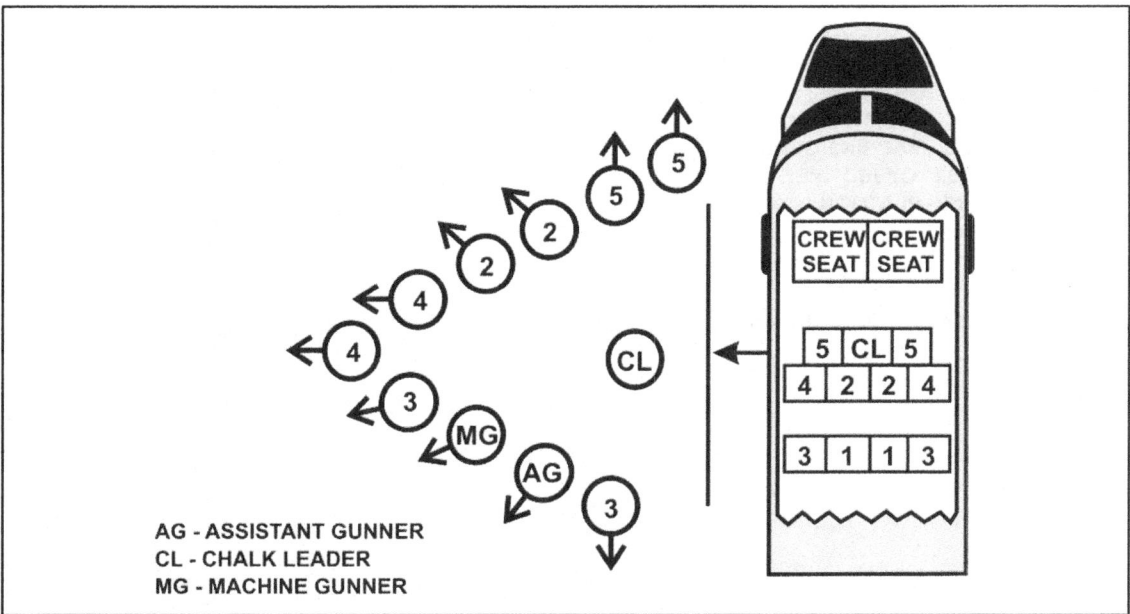

Figure 11-1. One-side off-load (UH-60)

Advantages

11-37. The one-side off-load simplifies mission command and the establishment of zones of responsibility on the landing zone. It allows the door gunners on the opposite side of the aircraft to engage enemy positions during off-loading. (See figure 11-2a, page 11-8.) This allows the door gunners of follow-on serials to engage enemy on the far side of the landing zone. Figure 11-2b on page 11-8 allows for immediate establishment of 360-degree security upon landing.

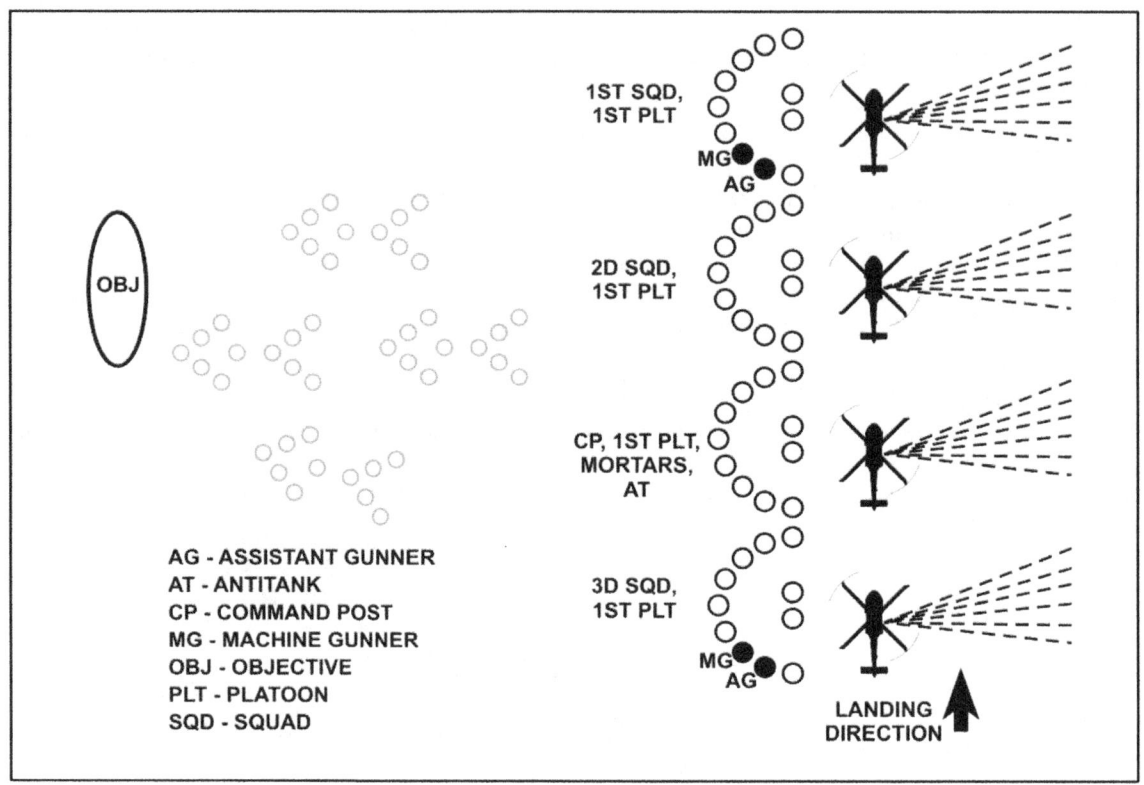

Figure 11-2a. One-side off-load (squads in same chalk) trail landing formation

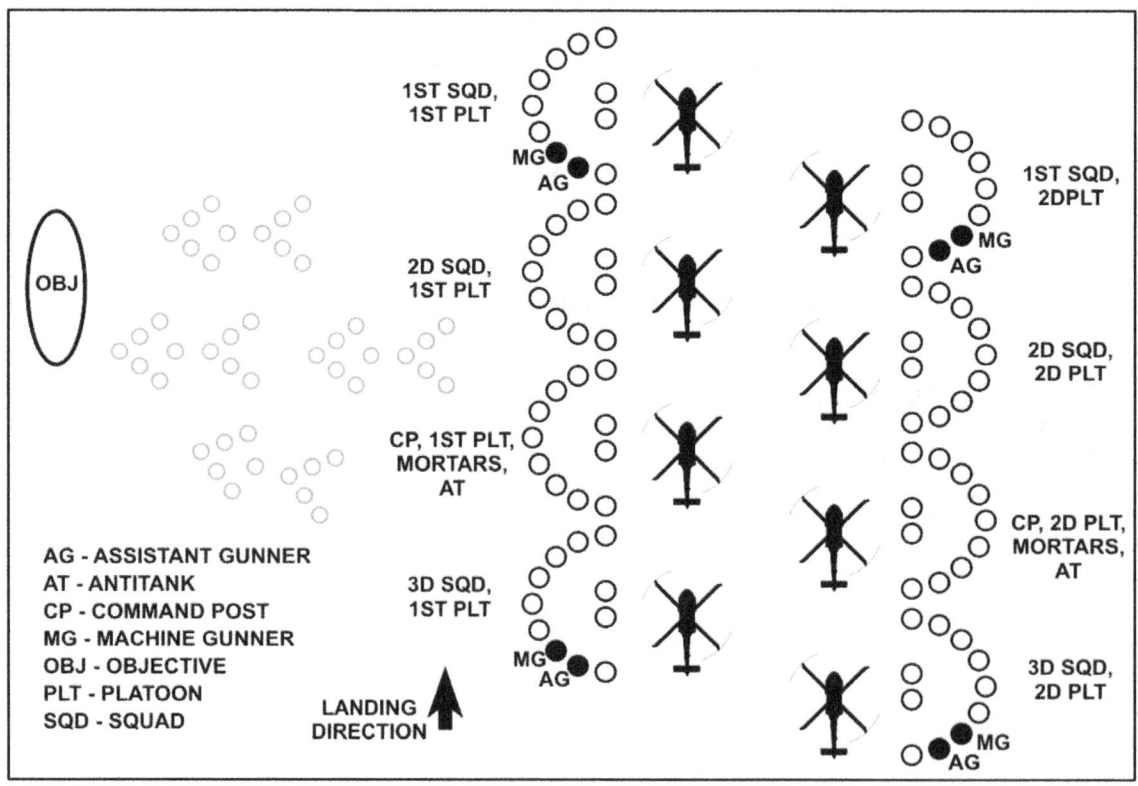

Figure 11-2b. One-side off-load (squads in same chalk) staggered trail right landing formation

Landing Plan

Disadvantages

11-38. The one-side off-load is the slowest of the off-loading methods. The Soldiers and aircraft are exposed for a longer amount of time while exiting the aircraft, making them vulnerable to direct and indirect fire.

TWO-SIDE OFF-LOAD

11-39. In this method, Soldiers exit from both sides of the aircraft. (See figure 11-3.) Soldiers exiting the aircraft should step outward and take up a prone position, forming 180-degree security on that side of the aircraft yet remaining under the main rotor system and outside the landing gear of the aircraft. Soldiers should remain in the prone position until the aircraft lifts off before departing the landing zone. The squad leader directs his squad to move directly to the nearest covered and concealed position according to the landing plan or SOPs.

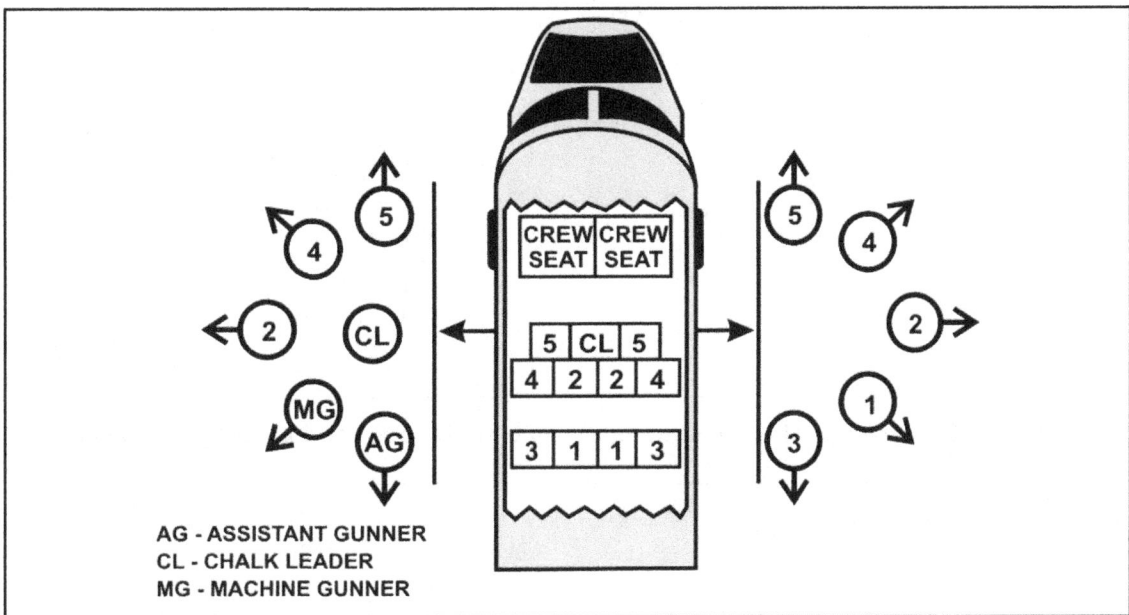

Figure 11-3. Two-side off-load (UH-60)

11-40. Cross-load options allow for pure unit integrity of chalks (See figure 11-4, page 11-10.) or mixed loads to support moving to opposite sides of a large pickup zone. (See figure 11-5, page 11-10.) Cross-load planning considerations support the mission command initially required on the landing zone and follow-on lifts into the landing zone.

Advantages

11-41. The two-side off-load is the quickest method for exiting the aircraft. It simplifies control and the establishment of zones of responsibility on the landing zone.

Disadvantages

11-42. The two-side off-load has the slowest movement time off the landing zone of all off-loading methods, which exposes Soldiers longer to enemy direct and indirect fire. This method masks both door gunner fires while Soldiers exit the aircraft, which increases vulnerability to enemy direct fire.

Chapter 11

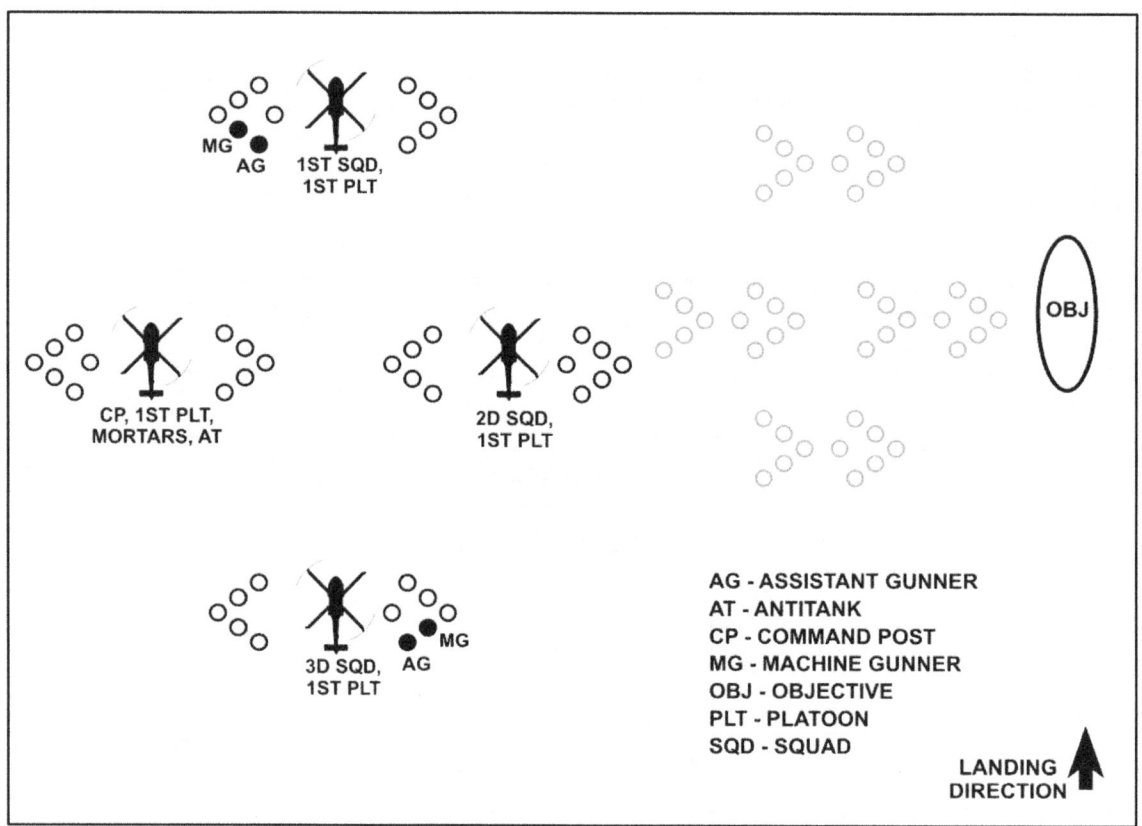

Figure 11-4. Two-side off-load (squads in same chalk) diamond landing formation

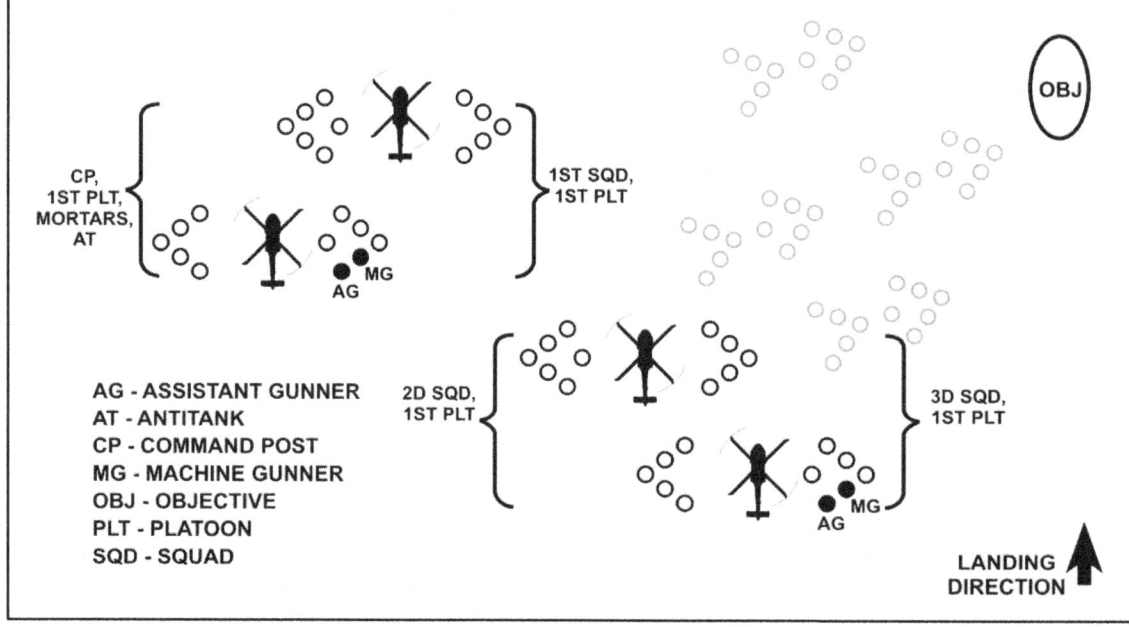

Figure 11-5. Two-side off-load (chalks cross-loaded) heavy right landing formation

REAR RAMP OFF-LOAD

11-43. In this method, Soldiers exit from the rear ramp of a CH-47 or other rear exiting aircraft. Soldiers move out from the aircraft and drop to a prone fighting position, establishing 360-degree security until the aircraft lifts to depart the landing zone. (See figure 11-6.) Once the aircraft departs the landing zone, the unit may execute a one- or two-side landing zone rush according to the landing plan or SOPs.

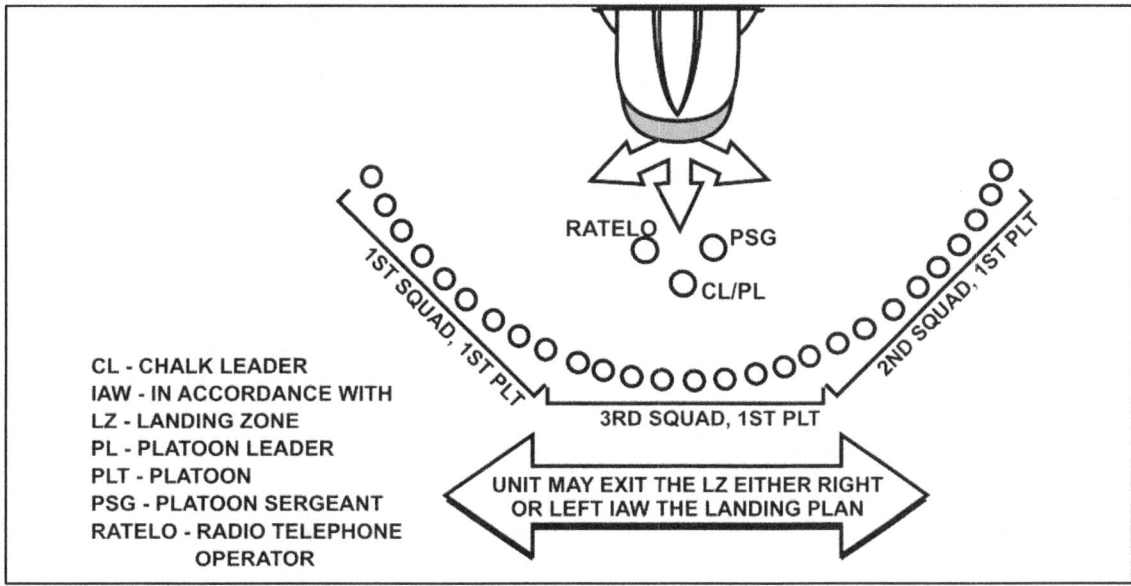

Figure 11-6. Rear ramp off-load and landing zone exit (CH-47)

EXITING THE LANDING ZONE

ONE-SIDE LANDING ZONE RUSH

11-44. Upon exiting the aircraft and dropping to the prone position, Soldiers recover from the prone position and move immediately with their squad to a covered and concealed position (such as a tree line) in wedge or other formation determined by their squad leader. Squads assemble at designated rally points and then move to assault objectives on the landing zone or to objectives off the landing zone. This is the preferred method to use when touchdown points are near covered and concealed positions. The unit may plan a one-side landing zone rush away from a potential enemy position, allowing the door gunner closest to the enemy position to continue firing while Soldiers exit the other side of the aircraft. (See figure 11-7, page 11-12.)

Chapter 11

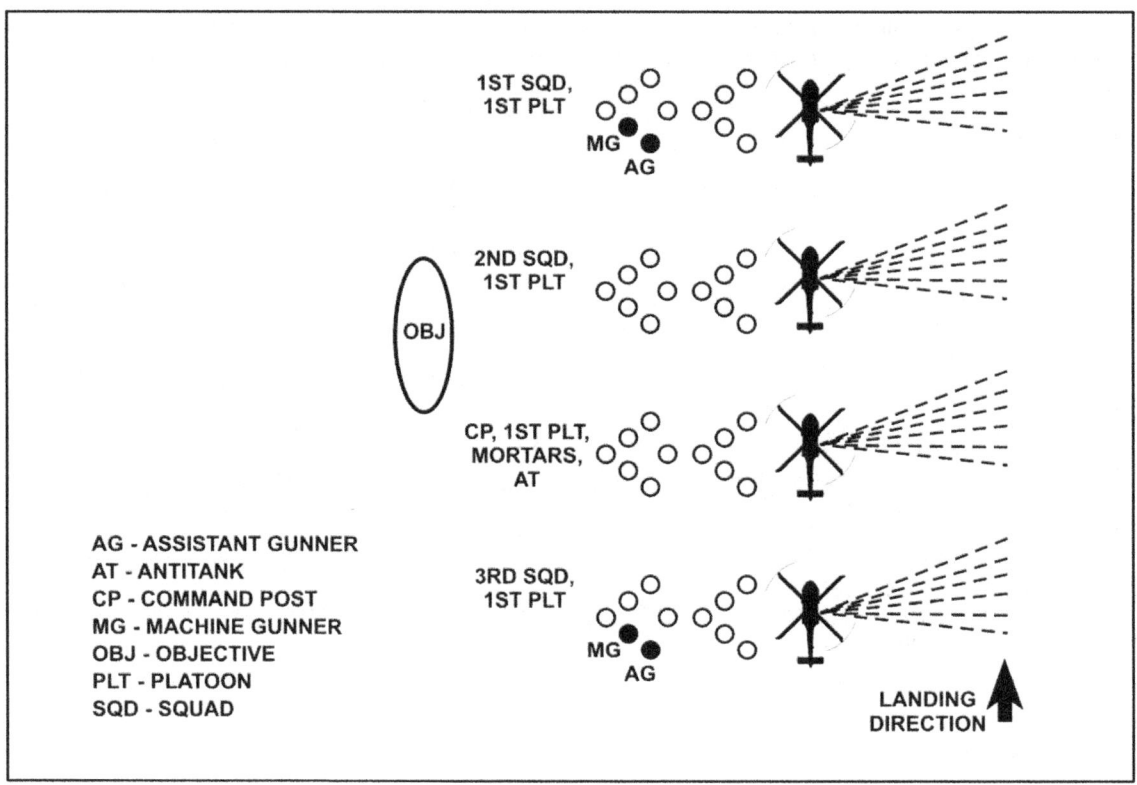

Figure 11-7. One-side landing zone rush (squads in same chalk) trail landing formation

Advantages

11-45. A one-side landing zone rush—
- Moves the unit off the danger area quickly.
- Facilitates control.
- Maintains momentum and is less vulnerable to indirect fires.
- Simplifies establishing zones of responsibility on the landing zone.
- Minimizes aircraft cross-loading plans.
- Allows door gunner of off-loading and follow-on serials to engage enemy on the far side of the landing zone.
- Clears the landing zone quickly for follow-on lifts.

Disadvantages

11-46. The unit executing a one-side landing zone rush is vulnerable to direct fire weapons while moving off the landing zone.

Two-Side Landing Zone Rush

11-47. Aircraft loading options to consider when using a two-side landing zone rush are:
- Split the squad across two chalks, with each fire team exiting the same door. (See figure 11-8.)
- Keep each chalk as a pure squad, with even-numbered chalks exiting the right door and odd-numbered chalks exiting the left door or vice versa. (See figure 11-9, page 11-13.)

11-48. Upon exiting the aircraft and dropping to the prone position, Soldiers recover from the prone position and move immediately with their squad to a covered and concealed position in wedge or other formation designated by their squad leader. Squads assemble at designated rally points and then move to assault objectives on the landing zone or to objectives off the landing zone. The aircraft landing formation can help facilitate the unit in rapidly clearing Soldiers off the landing zone.

Advantages

11-49. A two-side landing zone rush—
- Moves the unit off the danger area fastest.
- Facilitates clearing and securing of the landing zone.
- Facilitates fire control measures on the landing zone.
- Maintains momentum and is less vulnerable to indirect fires.
- Establishes zones of responsibility on the landing zone.
- Clears the landing zone quickly for follow-on lifts.

Disadvantages

11-50. A two-side landing zone rush is more difficult to plan and control due to its complex aircraft cross-loading plan. It masks fires of both door gunners while departing the landing zone, which increases vulnerability to direct fire while moving off the landing zone.

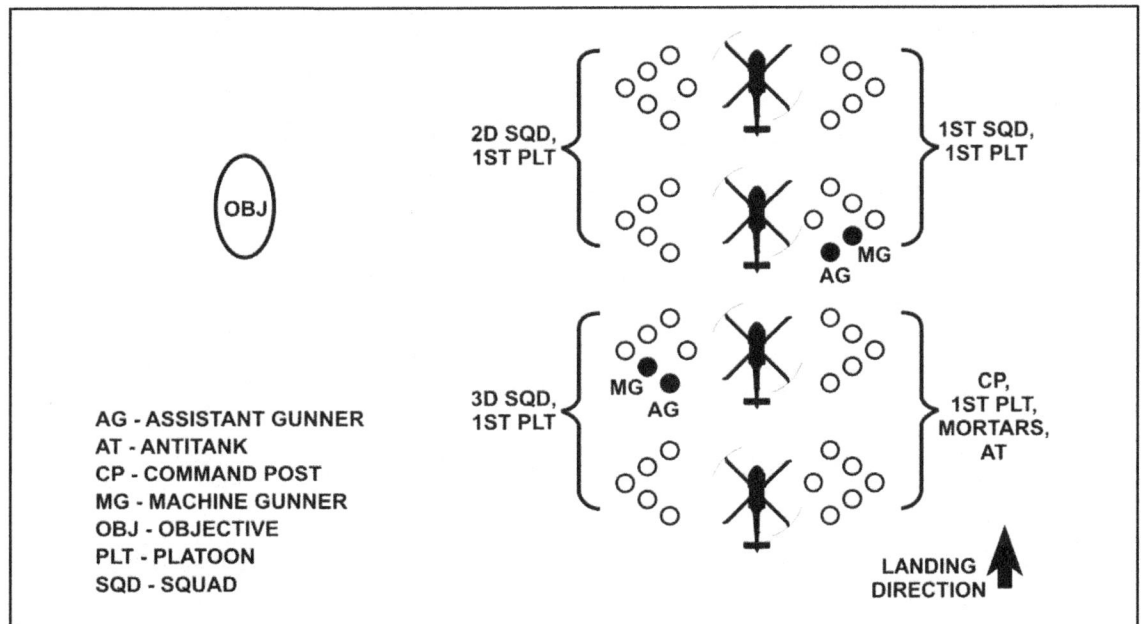

Figure 11-8. Two-side landing zone rush (chalks cross loaded) trail landing formation

Chapter 11

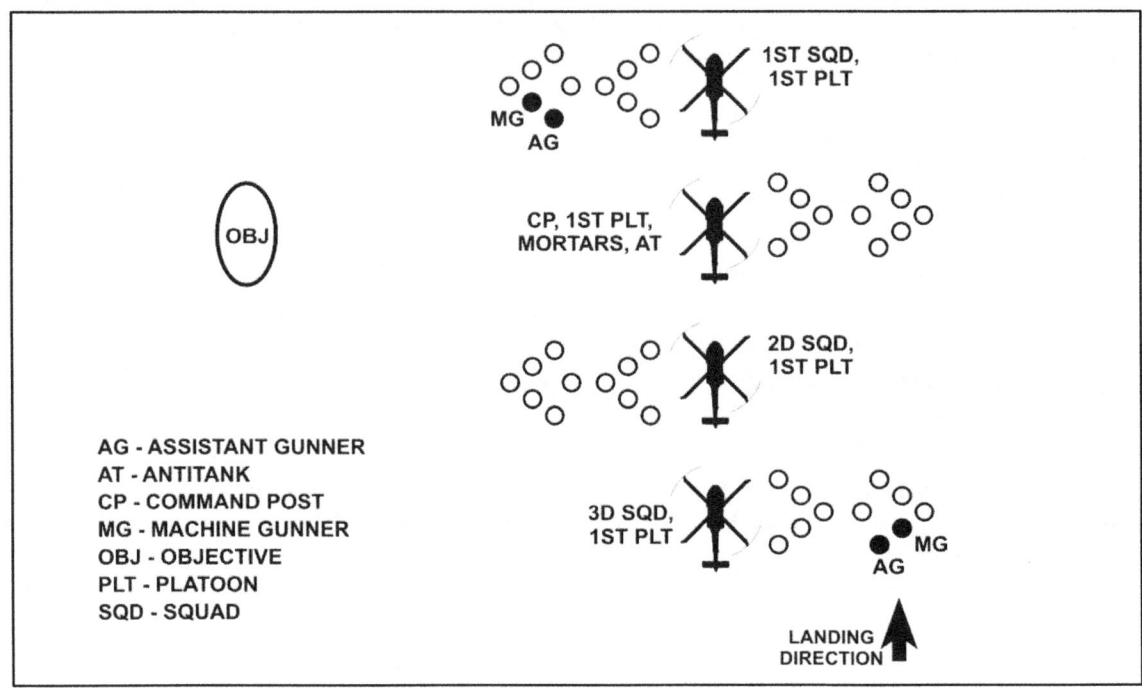

Figure 11-9. Two-side landing zone rush (squads in same chalk) trail landing formation

Chapter 12
Air Movement Plan

The air movement plan is largely based on the ground tactical plan and landing plan. It begins when the assault or lift helicopters cross the start point and ends when they cross the release point. The air movement plan specifies the schedule and provides instructions for air movement of Soldiers, equipment, and supplies from the pickup zone to the landing zone. The air movement plan considers the impact of airspace restrictions. It provides coordinating instructions regarding air routes, aircraft speeds, altitudes, formations, and the planned use of attack reconnaissance helicopters.

SECTION I – DEVELOPMENT CONSIDERATIONS

12-1. The air movement plan is developed by the air assault task force (AATF) and supporting aviation unit staffs in coordination with technical assistance and recommendations from the brigade aviation element, air mission commander, and the aviation liaison officer. The aviation unit conducts all air mission planning using the Aviation Mission Planning System (AMPS). This allows the aviation unit to plan digitally, allowing rapid distribution of digital products between units within the AATF. However, the air assault task force commander (AATFC) approves the final plan. The result of air movement planning is the completion of the air movement table, which specifies the AATF movement from the pickup zone to the landing zone.

12-2. Important considerations when developing the air movement plan are—
- Air routes.
- En route formations.
- Terrain flight modes.
- Fires.
- Suppression of enemy air defense.
- Air assault security.
- Mission command.

AIR ROUTES

12-3. Components of an air route are—
- Start point.
- Release point.
- Air control points.
- Flight path between the start point and release point.

START POINT AND RELEASE POINT

12-4. The air route starts at the start point and ends at the release point. The location of start points and release points are usually three to five kilometers from the pickup zones and landing zones respectively to allow adequate flying time for execution of the flight's en route procedures. The distance from the pickup zone to the start point allows the aircraft to achieve the desired airspeed, altitude, and formation after liftoff. The distance from the release point to the landing zone allows the flight leader to reconfigure the formation and execute a tactical formation landing. The designated locations of the start points and release points should—

- Profit from favorable weather conditions.
- Avoid obstacles and known enemy positions.
- Facilitate takeoff and landing into the wind by the best air route.

AIR CONTROL POINTS

12-5. Air control points designate each point where the air route changes direction. They include readily identifiable topographic features or points marked by electronic navigational aids. A route may have as many air control points as needed to control the air movement. The start points and release points are air control points.

12-6. Once identified, air routes are designated for use by each unit. When large groups of aircraft are employed, dispersion is achieved by using multiple routes. However, with large serials, it is often necessary to use fewer routes or even a single route to concentrate available supporting fires. The number of alternate and return routes may be limited.

CRITERIA FOR SELECTING ROUTES

12-7. Regardless of direction or location, certain criteria apply. All characteristics are seldom present in any one situation, but all should be considered. Give careful consideration to the terrain and enemy forces. Air routes should assist in navigation (day or night) and avoid turns in excess of 60 degrees to facilitate control of the aircraft formation when formation flying is required or if sling loads are involved.

12-8. Factors to consider when selecting routes as follows:
- Interference with ground action. Overflying ground elements may interfere with their supporting fire. Clear air routes of the gun-target line when possible. Avoid over-flight of built-up areas.
- Support of landing plan. To reduce vulnerability of the air assault force, air routes facilitate rapid approach, landing, and departure from selected landing zones.
- Enemy ground and air capabilities. Air routes maximize use of terrain, cover, and concealment to minimize exposure to enemy observation, target acquisition, and direct fire.
- Available fire support. Air routes allow fire support from all available resources. Avoid masking friendly fires, particularly supporting fires.
- Available air cover. Air routes are selected to provide air cover for friendly forces en route.
- Weather conditions. Prevailing weather during the air assault operation significantly affects the selection of air routes.
- Terrain. Air routes use terrain to maximize the advantage of and reduce vulnerability of the aircraft formations, providing cover by placing terrain mass and vegetation between the enemy and the aircraft.
- Distance from pickup zone to landing zone. Air routes should be as short as is tactically feasible according to mission variables to reduce flying time.

12-9. Maps or overlays containing air route information are prepared at aviation unit headquarters and disseminated to subordinate and support units. Air routes and corridors are designated by a letter, number, or word. (See figure 12-1, page 12-3.)

Air Movement Plan

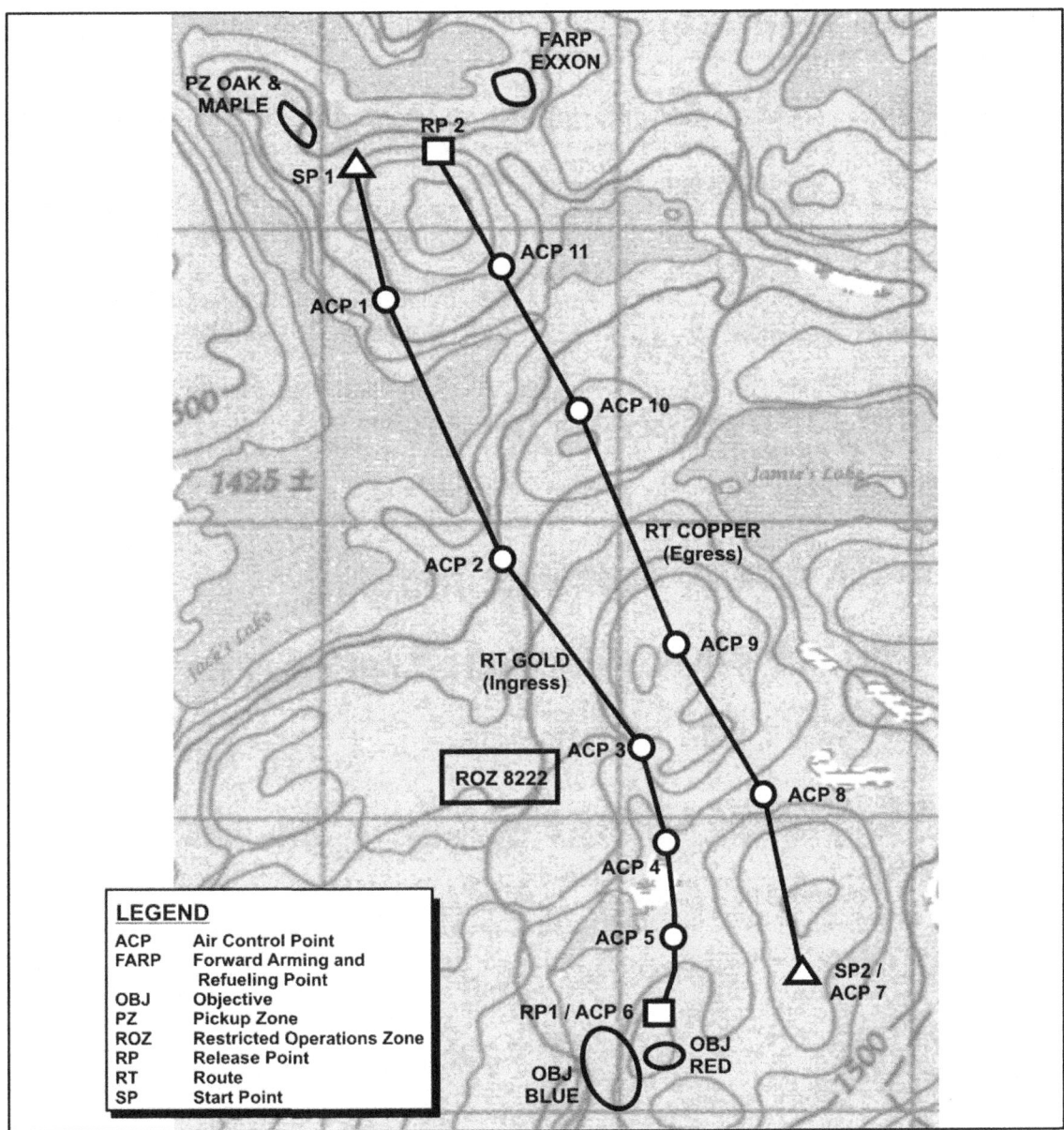

Figure 12-1. Air route overlay

EN ROUTE FORMATIONS

12-10. Many factors dictate the flight's formation, such as terrain, enemy situation, visibility, weather, altitude, speed, type of aircraft mix, and the degree of control required. The air mission commander or flight leader selects the en route formation and landing formation based on the mission analysis of the ground tactical plan. Ideally, all aircrafts land at the same time in a planned flight formation as specified by the air movement table. The landing site commander includes this information in his landing instructions to the flight leader and the pathfinder establishing the landing zone. (Refer to FM 3-04.113 for more information.)

12-11. The flight leader and pathfinder must understand the en route and landing formation and the ground tactical plan to best support the ground unit and facilitate the air assault operation. The flight leader should try to match the landing formation to the flight formation. Pilots should have to modify their formations no more than necessary to accommodate the restrictions of a landing site, but it might be necessary to land in a restrictive area. Touchdown points are established by the pathfinder in the same order as indicated in the formation. The following standard flight and landing formations (See figure 12-2, page 12-5.) are used when conducting air assault operations:

- Heavy left or right formation. Requires a relatively long, wide landing area; presents difficulty in pre-positioning loads; restricts suppressive fire by inboard gunners; provides firepower to front and flank.
- Diamond formation. Allows rapid deployment for all-round security; requires relatively small landing area; presents some difficulty in pre-positioning loads; restricts suppressive fire of inboard gunners.
- Vee formation. Requires a relatively small landing area; allows rapid deployment of forces to the front; restricts suppressive fire of inboard gunners; presents some difficulty in prepositioning loads.
- Echelon left or right formation. Requires a relatively long, wide landing area; presents some difficulty in prepositioning loads; allows rapid deployment of forces to the flank; allows unrestricted suppressive fire by gunners.
- Trail formation. Requires a relatively small landing area; allows rapid deployment of forces to the flank; simplifies pre-positioning loads; allows unrestricted suppressive fire by gunners.
- Staggered trail left or right formation. Requires a relatively long, wide landing area; simplifies pre-positioning loads; allows rapid deployment for all-round security; gunners' suppressive fire restricted somewhat.

Air Movement Plan

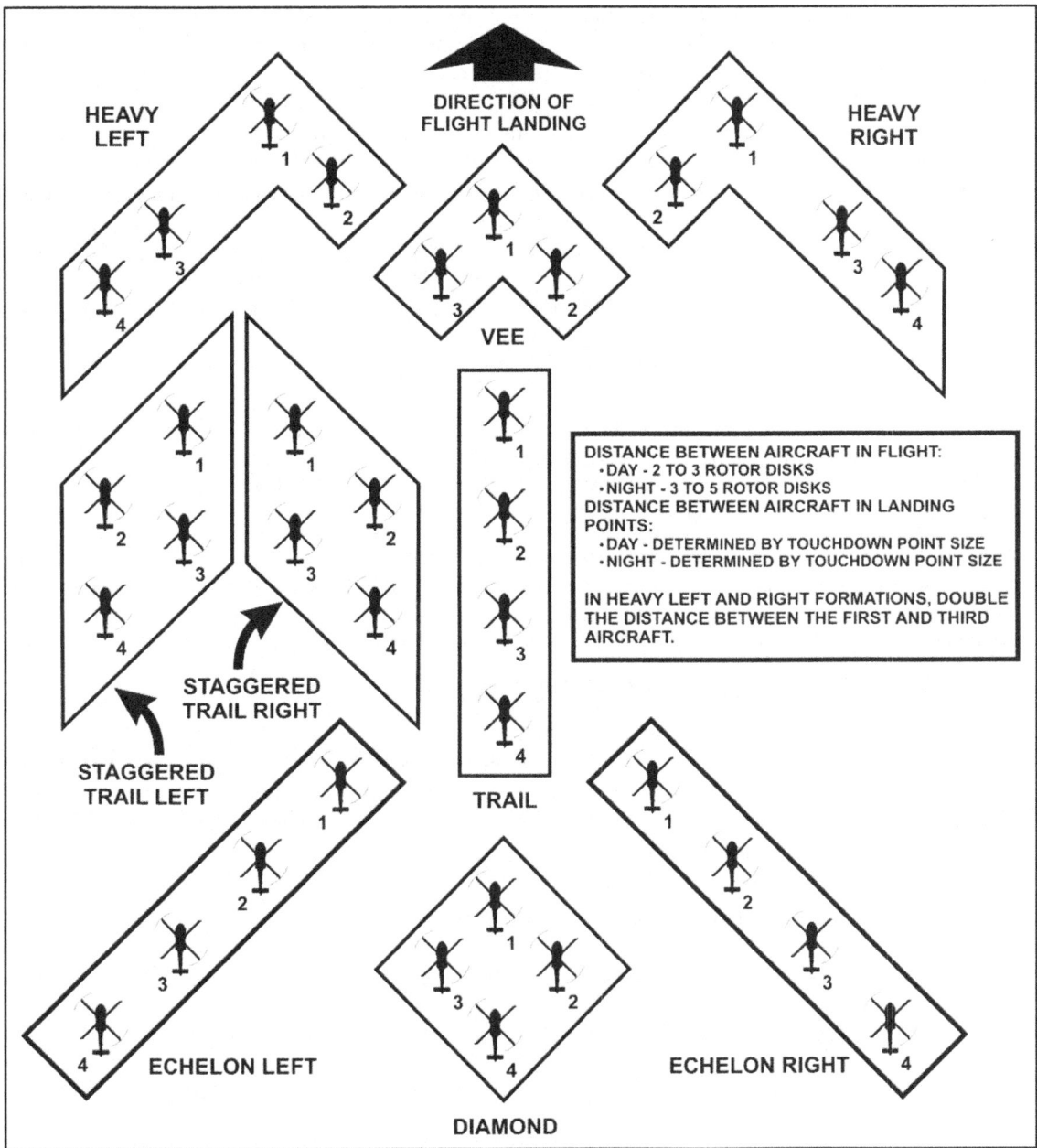

Figure 12-2. Standard flight and landing formations

12-12. The pathfinder chooses landing sites that have firm surfaces; are free of dust, sand, and debris that might create problems when disturbed by rotor wash; and are cleared of obstacles. The landing site is laid out in a location where helicopters will not fly directly over aircraft on the ground. The layout of the site also depends on the landing space available, the number and type of obstacles, unit standard operating procedures, and prearranged flight formations. En route formation impacts how the formation lands, impacting loading and off-loading of aircraft. (Refer to FM 3-21.38 for more information.)

TERRAIN FLIGHT MODES

12-13. A specific en route flight altitude is not designated and is usually below the coordinating altitude. Factors affecting flight altitude include enemy, terrain, navigation, weather, flight distance, need for surprise, and pilot fatigue. Pilots may use one or some combination of the three terrain flight modes as dictated by the mission variables.

- Nap-of-the-earth flight is conducted at varying airspeeds as close to the earth's surface as vegetation and obstacles permit. A weaving flight path remains oriented along the general axis of movement and takes advantage of terrain masking. This is a general flight mode and may likely be in close proximity to the enemy.
- Contour flight is conducted at low altitudes, conforming to the earth's contours. Relatively constant airspeeds and varying altitudes as dictated by terrain and obstacles characterize it.
- Low-level flight is conducted at constant altitudes and airspeed dictated by threat avoidance. Its intent is to facilitate speed and ease of movement while minimizing detection. This mode of flight is used when there is a low threat level. Fires along the air route are planned to suppress known or suspected enemy positions. These fires should be intense and of short duration. Utilize multiple target engagement methods as needed. On-call fires are planned along the air route to ensure rapid target engagement if necessary.

FIRES

12-14. Fire plans cover the pickup zones, air routes, and landing zones. Fire support plans include suppression of enemy air defenses (SEAD) systems and obscuration to protect formations from enemy detection. This requires aggressive fire planning and direct coordination with field artillery and mortar fire direction centers and other fire support elements.

12-15. All available fire support is used to suppress or destroy enemy weapons, to include close air support, artillery, and attack reconnaissance helicopters. Support may comprise concealment or other countermeasures for suppressing or confusing enemy air defense systems. During night operations, the use of illumination fire requires detailed planning. Illumination can interfere with NVGs causing unsafe conditions.

SUPPRESSION OF ENEMY AIR DEFENSES

12-16. In executing air movement, the air mission commander integrates air routes based on pickup zone and landing zone locations, avoiding known or suspected enemy air defense positions. The AATF is responsible for planning, synchronizing, and executing lethal suppressive fires and nonlethal suppressive effects on known or suspected enemy air defense positions that are unavoidable. Lethal and nonlethal assets available to conduct SEAD missions include—
- Mortars and artillery (cannon, rocket and missile).
- Fixed-wing assets, to include unmanned aircraft systems.
- Naval gunfire.
- Attack reconnaissance helicopters.
- Radar suppression and jamming (lethal and nonlethal).
- Communications suppression and jamming (lethal and nonlethal).

JOINT SEAD

12-17. The term Joint SEAD encompasses all SEAD activities provided by components of a joint force in support of one another. When operating as a component of a joint force, different assets and unique planning requirements may exist. (Refer to JP 3-01 for more information.) Joint SEAD includes all SEAD categories and additional classifications to include—
- Operational area system suppression comprises operations within an operational area against specific enemy air defense systems to degrade or destroy their effectiveness. It targets high payoff air defense systems whose degradation most affects the enemy's total system.
- Opportune suppression is a continuous operation involving immediate attack of air defense targets of opportunity. It is normally unplanned suppression, includes aircrew self-defense, and attacks against targets of opportunity.
- Localized suppression can occur throughout the area of responsibility or joint operations area and can be conducted by all components. However, it is limited in time and to geographical areas associated with specific ground targets.
- Corridor suppression is planned joint SEAD focused on creating an air defense artillery suppressed corridor to maneuver aircraft. Missions that normally require this suppression are air

missions supporting tactical airlift or combat operations, search and rescue operations, and operations in support of special operations forces.

SEAD PLANNING

12-18. The ground maneuver, aviation units, AATF operations officers, AATF intelligence officers (Refer to FM 2-0 for more information.), and electronic warfare officer (See ATP 3-36.) participate in SEAD planning. SEAD planning is conducted as part of the military decisionmaking process and targeting process. Consider the following critical factors in mission analysis:
- Ingress and egress air routes and locations of air control points.
- En route airspeed.
- Time, distance, and heading information for primary and alternate air routes.
- Expected start point crossing time on ingress and egress.
- Enemy air defense artillery locations within the area of operation.
- Locations, frequencies, and call signs of friendly artillery.
- Available assets to deliver SEAD fires.

12-19. When determining enemy air defense capabilities, mission planners—
- Plot the location of all known enemy air defense artillery systems on a map.
- Draw a circle (threat ring) around each air defense artillery system with a radius equal to the maximum engagement range. Depending on the threat system and its means of target acquisition (optical, infrared, and radar) and fire control, the size of the threat ring may change during hours of limited visibility. Terrain that blocks electronic or visual lines of sight may reduce the radius of a threat ring.
- Use AMPS, Falcon View, or other automated systems to reduce workload and ensure accuracy.
- Plot the primary and alternate air routes and all landing zones on the map. Air routes and landing zones should avoid threat rings whenever possible.

12-20. Plan SEAD fires to engage the two types of targets, planned targets and targets of opportunity described below.

Planned Targets

12-21. A *planned target* is a target that is known to exist in the operational environment, upon which actions are planned using deliberate targeting, creating effects which support the commander's objectives (JP 3-60). The two types of planned targets are—
- Scheduled targets that are prosecuted at a specified time.
- On-call targets that have planned actions and are triggered when detected or located.

12-22. One example of a scheduled target is a deception SEAD mission. Deception SEAD may be fired into an area to deceive the enemy or cause him to reposition his air defense weapons away from where actual operations take place. Another example is an electronic attack of enemy air defense radars and command and control information systems when enemy ADA assets are in civilian populated areas.

12-23. Provisions should also exist for immediate on-call fires in the SEAD plan. Establish a quick-fire network for this purpose providing a direct link between an observer and weapon system (normally field artillery). Order observers based on their priority of fire. Conduct a fire support rehearsal with the supporting unit. Brief and rehearse with all participants during the combined arms rehearsal.

Targets of Opportunity

12-24. SEAD is conducted against ADA targets of opportunity and should reflect priorities established on the high-payoff target list and attack guidance matrix. Delivery systems and quick-fire networks are critical to engaging targets of opportunity.

Chapter 12

SEAD EMPLOYMENT

12-25. SEAD fires should be planned against an enemy ADA system that threatens the air assault force. A period of focused immediate SEAD is planned at each landing zone before the arrival of the AATF. If possible, plan deception SEAD to mitigate further tactical risk.

12-26. Scheduled SEAD missions are planned against threat systems along the ingress and egress route of flight. The start time for each SEAD mission may be calculated if the assault aircraft's en route airspeed and SP time on the air route are known. These calculations may be made manually or with AMPS or similar planning systems.

12-27. Factors that determine the duration of each SEAD mission include aircraft speed and the range of each enemy ADA system (size of the threat ring). This information may be used with planning software to determine how long to suppress each ADA system along the air route. Calculations may be made manually or estimated. A good planning estimate is that the air assault travels three kilometers in one minute.

12-28. Position units to support as much of the area of operation as possible. To ensure synchronization, organize all planned fires into an SEAD schedule or add them to the execution matrix. Assess the effectiveness of the SEAD plan during war-gaming.

AIR ASSAULT SECURITY

12-29. Air assault security is conducted throughout the air movement phase. Air assault security is not necessarily just an escort mission. The air assault security process can be conducted sequentially, simultaneously, or over a period of 24 to 72 hours before the start of the air assault mission. This process is determined early in the mission analysis phase and is a direct result of the AATFC's initial guidance and key tasks.

12-30. UAS should observe the air routes and landing zones beginning well before launch to provide early warning to the AATFC. Just before the launch of the air movement phase, attack reconnaissance units fly along the route to conduct an air assault security mission. This mission is much like a movement to contact. Usually, one to two attack reconnaissance companies conduct the mission just before the assault aircraft launch for the air movement. This allows the attack reconnaissance units opportunity to conduct a relief on station with elements that may already be on station providing reconnaissance.

12-31. The air assault security force generally makes the final landing zone update call, as the assault forces are en route to the landing zone. Before assault forces land on the landing zone, air assault security forces may be directed to shift to a landing zone overwatch mission, ensuring they do not conflict with the air routes entering or exiting the landing zone. As the assault forces land on the landing zone, air assault security forces may be directed to move forward to the next phase line to conduct a screening mission or to occupy a battle position.

12-32. Attack reconnaissance units maintain the flexibility to execute on-call close combat attacks as needed. Air assault security forces must maintain communications with the fires elements for immediate suppression missions as needed.

MISSION COMMAND

12-33. In executing the air movement, the air mission commander takes operational control of all Army aviation forces. The air mission commander controls all—
- Timing for deconfliction.
- En route fires.
- Initiation and shifting of landing zone preparation fires.

12-34. Once the air assault force has cleared the landing zone and moved to its rally point, the tactical commander on the ground assumes mission command of the element and continues his assigned mission. Mission command should allow continued execution despite loss of radio communications. If the air mission commander and lift flight leaders have air movement tables or the execution checklist in their possession, they can continue the mission without radio communications.

SECTION II – AIR MOVEMENT TABLE

12-35. The AATF staff and aviation unit staff develop the air movement table. (See table 12-1, page 12-10.) This table serves as the primary air movement document for the air assault operation.

AIR MOVEMENT TABLE DEVELOPMENT

12-36. The AATF S-3 Air and aviation liaison officer begin work on this document right after the initial planning conference. This gives them an idea early in the planning process of challenges involved in moving units to the landing zone. The table ensures that all personnel, equipment, and supplies are accounted for in the movement and that each aircraft is fully loaded, correctly positioned in the flight, and directed to the right landing zone. The air movement table—
- Contains aircraft allocations.
- Designates number and type of aircraft in each serial.
- Specifies departure point; route to and from loading area; and loading, liftoff, and landing times.
- Includes the refuel schedule for all lifts if required.

AIR MOVEMENT TABLE CRITERIA

12-37. The air movement table regulates the sequence of flight operations from pickup zone to landing zone using the following line information:
- Line number. Quick reference with brevity codes numbered sequentially.
- Aviation unit. Aviation unit conducting the air movement. Depicted as unit designation over call sign to save space.
- Lifted unit. Unit being lifted or air assaulted. If more than one unit is in the load, use unit with most assets in the load. Depicted as unit's designation over call sign to save space.
- Lift. Serials that make one complete turn out to and back from the area of operation. Numbered sequentially.
- Serial. A tactical grouping of two or more aircraft under the control of a serial commander (aviator) and separated from other tactical groupings within the lift by time or space. The capacity of the smallest landing zone determines the number of aircraft in each serial.
- Chalk. Each aircraft equals one load. Number UH-60 and CH-47 chalks separately.
- Pickup zone. Name of the pickup zone where chalks pick up the loads.
- Pickup zone arrival and load time. Time the troops get on the aircraft or when the aircraft starts to hookup the load.
- Takeoff time. Time the aircraft lifts off the pickup zone.
- Start point time. Time the aircraft hit the start point (brigade aviation element-determined point usually three to five kilometers from the pickup zone).
- Release point time. Time the aircraft hit the release point (brigade aviation element-determined point usually three to five kilometers from the landing zone).
- Landing zone. Landing zone name and location determined by the lifted unit's ground tactical plan.
- Landing zone time. Time the serial lands in the landing zone.
- Landing zone degree. Compass heading at which the serial is landing, should be converted to and shown in magnetic heading for the aircraft.
- Landing zone formation. Landing formation, normally the trail formation.
- Routes. Primary ingress and egress routes for the mission.
- Load. Personnel and sling load configuration. Refer to the tadpole diagram to save space on this page.
- Remarks. Additional remarks (such as scheduled delays, refuel, or other uncommon serial characteristics).

Table 12-1. Example air movement table

Air movement table

Line #	Avn Unit	Lifted Unit	Lift #	Serial	Chalk	PZ	PZ Arr/ Load Time	T/O Time	SP Time	RP Time	LZ	LZ Time	LZ Hdg	LZ Form	Routes Ingress	Routes Egress	Load PAX	Load Sling	Remarks
1	4-379	SCT/1-603 IN				As per coord	As per coord				Raven	H-36+00:00	As per PIC	As per PIC	As per PIC	As per PIC	16		False into Lark. To recon Robin
2	4-354	SCT/2-603 IN				As per coord	As per coord				Oriole	H-35+59:00	As per PIC	As per PIC	As per PIC	As per PIC	16		False into Bluejay To recon Sparrow
3	3-354	2/C/6-4 CAV (-)				As per coord	As per coord				Pelican	H-35+58:00	As per PIC	As per PIC	As per PIC	As per PIC	20		False into Emu To recon Crow
4	3-354	2/C/6-4 CAV (-)				As per coord	As per coord				Dove	H-35+57:00	As per PIC	As per PIC	As per PIC	As per PIC	8		False into Cardinal To recon Eagle
5	2-344	A/6-4 CAV	1	1	1-5	Oak	H-3+00:00	H-48:40	H-44:21	H-02:22	Crow	H-Hour	134	TRL	Gold (Crow)	Silver (Crow)	6 per a/c (30)	10x M1151	Refuel FARP EXXON
6	2-344	A/6-4 CAV	1	2	6-9	Oak	H-3+00:00	H-47:40	H-43:21	H-01:22	Crow	H+01:00	134	TRL	Gold (Crow)	Silver (Crow)	6 per a/c (24)	8x M1151	Refuel FARP EXXON
7	2-344	C/6-4 CAV (-)	1	3	10-12	Oak	H-3+00:00	H-46:40	H-42:21	H-00:22	Crow	H+02:00	134	TRL	Gold (Crow)	Silver (Crow)	6 per a/c + 40 (52)	4x M998	Refuel FARP EXXON
8	4-379	A/1-603 IN	2	1	1-5	Maple	H-2+00:00	H-44:40	H-40:21	H+01:38	Robin	H+02:58	112	TRL	Gold (Robin)	Silver (Robin)	5x16 (80)		
9	4-379	A/1-603 IN, HHC/1-603 IN	2	2	6-9	Maple	H-2+00:00	H-43:40	H-39:21	H+02:38	Robin	H+03:58	112	TRL	Gold (Robin)	Silver (Robin)	4x16 (64)		Chalk 9 is the BN CP
10	4-379	B/1-603 IN	2	3	10-14	Maple	H-2+00:00	H-42:40	H-38:21	H+03:38	Robin	H+04:58	112	TRL	Gold (Robin)	Silver (Robin)	5x16 (80)		
11	4-379	B/1-603 IN, 1/E/603 EN	2	4	15-18	Maple	H-2+00:00	H-41:40	H-37:21	H+04:38	Robin	H+05:58	112	TRL	Gold (Robin)	Silver (Robin)	4x16 (64)		Chalk 18 is 1/E/603 EN
12	3-354	D/1-603 IN	3	1	13-16	Oak	H-3+00:00	H-40:40	H-36:21	H+05:38	Robin	H+06:58	112	TRL	Gold (Robin)	Silver (Robin)	8 per a/c (32)	8x M1151	Refuel FARP EXXON

Chapter 13
Loading and Staging

The activities that take place in or near the pickup zone are referred to as pickup zone operations. These activities include both the loading and staging plan. Like the previous steps in the air assault planning process, these plans support and are based on the steps before them. Pickup zone operations are a collaborative effort between the supported unit (maneuver forces that compose the assault force) and the supporting aviation unit. The assault force is organized on the pickup zone. Every serial and lift is a self-contained element that must understand what it does upon landing at either the primary or the alternate landing zone and later in executing the ground tactical plan. Planning for insertion and extraction follows the same process and requires the same forethought and attention to detail. Insertion and extraction plans are developed during the air assault planning process and coordinated with all supporting units at the initial planning conference or air mission coordination meeting (AMCM). Both insertion and extraction loading and staging plans should be rehearsed at the air assault task force (AATF), aviation, and assault force rehearsals.

SECTION I – LOADING PLAN

13-1. The loading plan ensures that Soldiers, equipment, and supplies are loaded on the correct aircraft and moved from the pickup zone to the landing zone in the priority order designated by the air assault task force commander (AATFC). The air movement table is the planning document that details how to execute this. At the company level and below, leaders use an air-loading table to document how the loading plan is executed. The basic information found in the air-loading table is found in the air movement table. Considerations to develop a loading plan are described below.

PICKUP ZONE SELECTION

13-2. Operations requiring the pick up or extraction of personnel may require special considerations dependent on the mission and or element requiring support. When the pickup zone or mission prevents a helicopter from landing, the mission may require the use of the special patrol infiltration and exfiltration system (SPIES). (Refer to FM 3-05.210 for more information.)

IDENTIFYING PICKUP ZONES

13-3. Identifying pickup zones is the first step in developing a loading plan. The goal of pickup zone identification is to locate suitable areas to accommodate the lift aircraft. Identify primary and alternate pickup zones at the same time.

PICKUP ZONE OPERATION

13-4. Establishing and running a pickup zone to standard is the first step in executing a successful air assault. The number of pickup zones selected depends on the number and type of aircraft and loads required to complete the mission. The mission may require the designation of both a light pickup zone (UH-60) and a heavy pickup zone (CH-47). Based on his unit's level of training, the air mission commander may adjust the specifications for identifying and selecting pickup zones (such as degree of slope, wind speeds, and distance between aircraft).

SELECTION CRITERIA

13-5. Once available pickup zones are identified, the AATFC and his S-3 select and assign pickup zones for each unit to use. Pickup zone selection criteria include:
- Number. Multiple pickup zones avoid concentrating forces in one area.
- Size. If possible, each pickup zone should accommodate all supporting aircraft at once.
- Proximity to Soldiers. When possible, the selected pickup zones should not require extensive ground movement to the pickup zone by troops.
- Accessibility. Each pickup zone should be accessible to vehicles to move support assets and assault forces.
- Vulnerability to attack. Selected pickup zones should be masked by terrain from enemy observation.
- Conditions. Surface conditions of the area (for example, excessive slope; blowing dust, sand, or snow; and man-made obstacles) create potential hazards to pickup zone operations.

Note. Using pickup zones located in secure forward operating base and outposts precludes much of the effort required to identify and select suitable pickup zones.

PICKUP ZONE ORGANIZATION AND CONTROL

13-6. Once the AATFC selects the pickup zones, he designates a pickup zone control officer (PZCO) to organize, control, and coordinate pickup zone operations. The designated PZCO is selected based on experience and the size of unit that is conducting the air assault. For example, at BCT level, the BCT executive officer is usually the PZCO. At the battalion level, the battalion executive officer or S-3 Air are usually the PZCO. At company level, the company executive officer is usually the PZCO.

13-7. Once designated, the PZCO is responsible for the overall success of all pickup zone activities, to include the following:
- Forming a control party to establish control over the pickup zone by clearing the pickup zone and establishing pickup zone security. The pickup zone control party comprises pickup zone control teams and support personnel from subordinate units, typically to include a PZCO, a pickup zone noncommissioned officer in charge (PZNCOIC), and—
    - Chalk guides guide the aircraft loads (Soldiers, vehicles, and equipment) from the chalk check-in point to their respective staging areas on the pickup zone once they have been inspected and approved for loading by the pickup zone control party.
    - Ground crew teams provide visual guidance to the aircraft pilots and hook up the vehicles and equipment that are externally loaded (sling loaded) by the aircraft. UH-60 ground crew teams typically consist of one hook-up person, one static probe person, and a signal person. CH-47 hook-up teams typically consist of one hook-up person and one static probe person according to sling hook-up point.
    - Crisis action teams are experienced officers or non-commissioned officers who are experts with rigging all types of loads and hook-up procedures for all aircraft.
    - Security teams provide local security for all pickup zone operations. These teams may include air defense teams if they are available.
    - Air traffic control teams (if available) use radio or directional light signals to provide flight information, expedite traffic, and prevent collisions. Pathfinder teams are capable of serving as air traffic control teams if required.
    - Pathfinder teams (if available) provide air traffic advisories and navigational aid for fixed- and rotary-wing aircraft. They perform limited physical improvement and chemical, biological, radiological, and nuclear monitoring and surveying within pickup zones, if required. Pathfinder availability, the tactical plan, the complexity of the operation, the terrain, and the air assault proficiency of the supported ground force may dictate pathfinder support.

- Establishing communications on two primary radio frequencies— one to control movement and loading of units and the other on combat aviation net. Alternate frequencies are provided as needed.
- Planning and initiating fire support near pickup zones in coordination with the AATF to provide all-round protection (from available support) without endangering arrival and departure of Soldiers or aircraft.
- Planning and initiating security to protect the main body as it assembles, moves to the pickup zone, and is lifted out. Other forces should provide security elements if the pickup zone is within a friendly area. Security comes from AATF resources if a unit is to be extracted from the objective area.
- Marking the pickup zone as specified in unit standard operating procedure regardless of the type of markers, pickup zone marking requirements depend on the type and number of aircraft and are based on the minimum acceptable distance between aircraft. At a minimum, mark the pickup zone to indicate where each aircraft, by type, is to land.
- Clearing the pickup zone of obstacles.
- Executing the bump plan.

COORDINATION WITH SUPPORTING AVIATION UNIT

13-8. Loading plans are carefully coordinated with the brigade aviation officer and aviation liaison. Copies of the air movement tables and air loading tables should be distributed to the aviation liaison officer, AATFC air mission commander, and PZCO.

13-9. The supporting helicopter unit must ensure that aviation expertise is present on the pickup zone. The brigade aviation officer or aviation liaison officer (or another designated representative) should locate with the PZCO during the pickup zone selection, setup, and execution phase. The aviation representatives provide guidance on the pickup zone setup, considering aircraft factors. For example, the pickup zone landing direction may change if the wind changes significantly. Additionally, the aviation representatives can offer advice on surface conditions and their effects on helicopter operations.

PREPARATION OF AIR LOADING TABLES

13-10. The air-loading table assigns personnel and major items of equipment or supplies to a specific aircraft (chalk) at the company and below level. The air-loading table is an accountability tool, a loading manifest, for each aircraft. (See table 13-1)

Table 13-1. Example air loading table

LINE #	AVN UNIT	LIFTED UNIT	LIFT	SERIAL	CHALK	PZ	PZ ARR/ LOAD TIME	T/O TIME	SP	RP	LZ	LZ TIME	LOAD PAX	LOAD SLING	REMARKS
1	2-344	A/1-603 IN	2	1	1-4	MAPLE	H-2+00:00	H-44:40	H-40:21	H-01:38	ROBIN	H+02:58	4X11 (44)		
2	2-344	A/1-603 IN	2	2	5-8	MAPLE	H-2+00:00	H-43:40	H-39:21	H+02:38	ROBIN	H+03:58	4X11 (44)		
3	2-344	A/1-603 IN	2	3	9-12	MAPLE	H-2+00:00	H-42:40	H-38:21	H+03:38	ROBIN	H+04:58	4X11 (44)		
4	2-344	A/1-603 IN, HHC 1-603 IN	2	4	13-16	MAPLE	H-2+00:00	H-41:40	H-37:21	H+04:38	ROBIN	H+05:58	4X11 (44)		CHALK 13 IS THE TAC

LEGEND
ARR Arrival
AVN Aviation
HHC Headquarters and Headquarters Company
IN Infantry
LZ Landing Zone
PZ Pickup Zone
RP Release Point
SP Start Point
TAC Tactical Command Post
T/O Take Off

13-11. When time is limited, the table can be written on a sheet of paper. It should contain a list, prepared by the aircraft chalk leader, of Soldiers (by name) and equipment to be loaded on each chalk. This ensures that information on personnel and equipment onboard is available if an aircraft is lost. The chalk leader gives a copy of the air-loading table to the pickup zone control party upon arriving at the pickup zone for check-in.

13-12. During preparation of the loading tables, leaders at all levels maintain the—
- Tactical integrity of units. Load a complete tactical unit, such as a fire team or squad, on the same aircraft or a platoon in the same serial to ensure integrity as a fighting unit upon landing.
- Tactical cross loading. Plan loads so that key personnel and critical equipment (for example, crew-served weapons) are not loaded on the same aircraft. Thus, if an aircraft is lost to an abort or enemy action, the mission is not seriously hampered.
- Self-sufficiency of loads. Ensure that each unit load has everything required (weapons, crew, and ammunition) to be operational upon reaching its destination. Ensure the following:
  - The prime mover accompanies every towed item.
  - Crews are loaded with their vehicle or weapon systems.

13-13. Leaders must determine whether internal or external (sling) loading is the best delivery method for equipment and supplies. Helicopters loaded internally can fly faster and are more maneuverable. Helicopters loaded externally fly slower at higher altitudes and are less maneuverable but can be loaded and unloaded more rapidly than internally loaded helicopters. The method used depends largely on availability of sling loading and rigging equipment.

DISPOSITION OF LOADS ON PICKUP ZONE

13-14. Position personnel and equipment on the pickup according to the pickup zone diagram. (See figure 13-1.) Flight crews must understand the loading plan and should be prepared to accept Soldiers and equipment immediately on landing. Pickup zone diagrams depicting the location of chalks and sling loads in the pickup zone assist flight crews in loading troops and equipment quickly once the aircraft arrive in the pickup zone. Flight crews should be provided a pickup zone diagram.

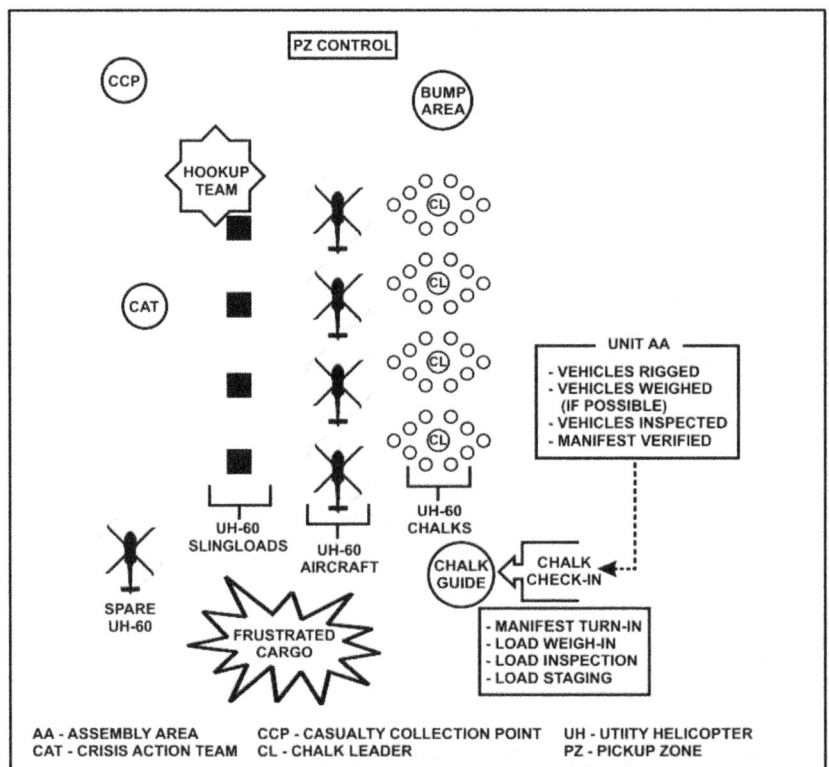

Figure 13-1. Example pickup zone diagram

LIFTS, SERIALS, AND CHALKS

13-15. The loading plan and pickup zone selection should aim to maintain ground unit integrity. Just as a squad should not be divided between chalks, a platoon should remain in one serial and a company should not be divided into different lifts or pickup zones. To maximize operational control, aviation assets are designated into lifts, serials, and chalks. (See figure 13-2, page 13-6.)

LIFTS

13-16. A lift is complete each time all aircraft assigned to the mission pick up Soldiers or equipment and set them down on the landing zone. The next lift is complete, when all lift aircraft place their next chalk on the landing zone and so on with all subsequent lifts.

SERIALS

13-17. A serial is a tactical grouping of two or more aircraft under the control of a serial commander (aviator) and separated from other tactical groupings within the lift by time or space. The use of serials may be necessary to maintain effective control of aviation assets. For example, due to METT-TC considerations, it may be difficult to control 16 aircraft as a single serial. However, a lift of 16 aircraft with four serials of four aircraft each can be more easily controlled.

13-18. Multiple serials may be necessary when the capacity of available pickup zones or landing zones is limited. If available pickup zones or landing zones can accommodate only four aircraft in a lift of 16 aircraft, it is best to organize into four serials of four aircraft each.

13-19. Multiple serials are employed to take advantage of available air routes. If several acceptable air routes are available, the AATFC may choose to employ serials to avoid concentrating his force along one air route. If the commander wants all his forces to land simultaneously in a single landing zone, he does so by having the serials converge at a common release point before landing. With a lift of 16 aircraft and four available air routes, the ABNAFC can use four serials of four aircraft each, with each serial using a different air route. Each time there is a new lift, a new serial begins. For example, within lift 1, there are serials 1 through 4. For each lift thereafter, serials start again with one.

CHALKS

13-20. A chalk comprises personnel and equipment designated to be moved by a specific aircraft. When planning the air movement, each aircraft within the lift is termed a chalk. For example, within a lift of 10, there are aircraft chalks 1 through 10. For each lift thereafter, there are chalks 1 through 10. Each aircraft is accounted for within each lift.

13-21. Chalks must be designated within serials just as they are within lifts. Counting within the serials is continuous up to the total number of aircraft in the lift. For example, in a lift of 16 aircraft in lift 1, serial 1, there are chalks 1 through 4. In lift 1, serial 2, there are chalks 5 through 8. In lift 1, serial 3, there are chalks 9 through 12. Finally, in lift 1, serial 4, there are chalks 13 through 16.

Chapter 13

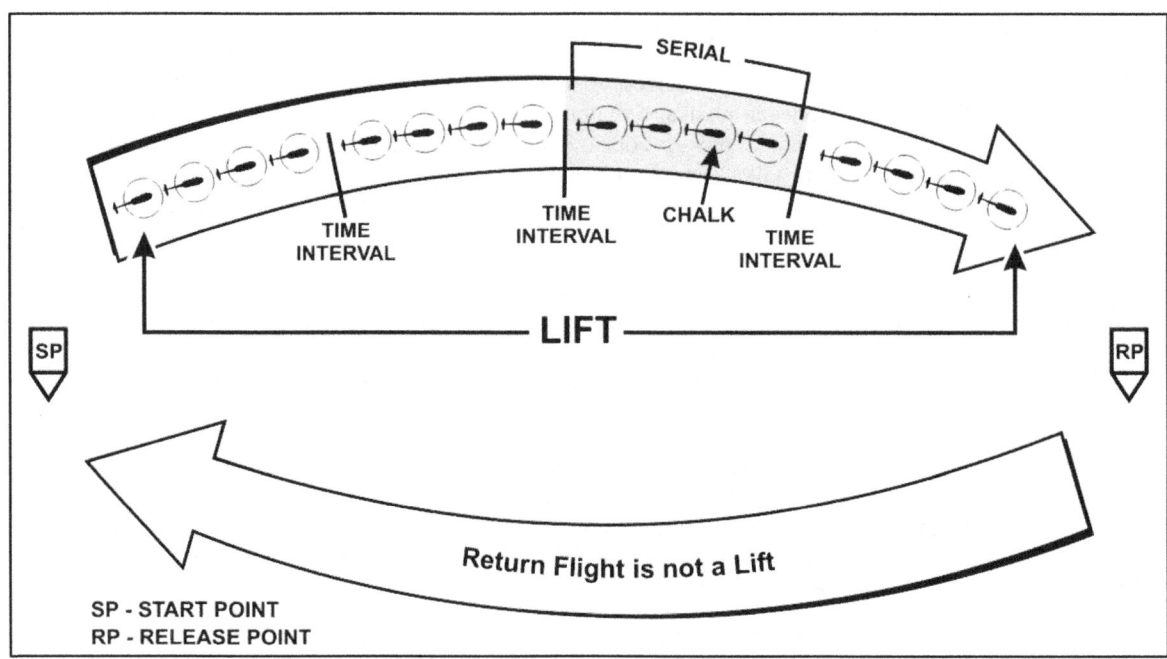

Figure 13-2. Lifts, serials, and chalks

BUMP PLAN

13-22. The bump plan ensures that the most essential personnel and equipment arrive on time at the objective area. It specifies personnel and equipment that may be bumped from an aircraft or serial, and delivered later. Each aircraft load and serial has a bump plan sequence designated on its air movement table. (See table 13-2.)

Table 13-2. Aircraft bump information

LINE #	AVN UNIT	LIFTED UNIT	LIFT	SERIAL	CHALK	PZ	PZ ARR/ LOAD TIME	T/O TIME	SP	RP	LZ	LZ TIME	LOAD PAX	LOAD SLING	REMARKS
1	2-344	A/1-603 IN	2	1	1-4	MAPLE	H-2+00:00	H-44:40	H-40:21	H-01:38	ROBIN	H+02:58	4X11 (44)		BUMP 4
2	2-344	A/1-603 IN	2	2	5-8	MAPLE	H-2+00:00	H-43:40	H-39:21	H+02:38	ROBIN	H+03:58	4X11 (44)		
3	2-344	A/1-603 IN	2	3	9-12	MAPLE	H-2+00:00	H-42:40	H-38:21	H+03:38	ROBIN	H+04:58	4X11 (44)		BUMP 12
4	2-344	A/1-603 IN, HHC 1-603 IN	2	4	13-16	MAPLE	H-2+00:00	H-41:40	H-37:21	H+04:38	ROBIN	H+05:58	4X11 (44)		CHALK 13 IS THE TAC CP

LEGEND
ARR   Arrival   RP   Release Point
AVN   Aviation   SP   Start Point
HHC   Headquarters and Headquarters Company   TAC   Tactical Command Post
IN   Infantry   T/O   Take Off
LZ   Landing Zone
PZ   Pickup Zone

13-23. If all personnel within the chalk cannot be lifted, individuals must know who is to offload and in what sequence. This ensures that key personnel are not bumped arbitrarily. This ensures that key aircraft chalks are not left in the pickup zone. When an aircraft within a serial or flight cannot lift off and key personnel are onboard, they offload and board another aircraft that has priority.

13-24. Bumped personnel report to a pickup zone bump area specified by company or larger units. At this location, they are accounted for, regrouped, and rescheduled by the PZCO for later delivery to appropriate landing zones. Sometimes, spare aircraft are held in reserve for bumped chalks in the event a primary mission aircraft is unable to fly due to maintenance or other reasons. These spare aircraft remain staged on the pickup zone for occasions such as these or to fly other high priority serials.

SECTION II – STAGING PLAN

13-25. The staging plan organizes the movement of Soldiers and loads into position for the forthcoming air assault. It establishes the pickup zone and specifies the manner in which the supported unit organizes to execute the loading plan. The staging plan prescribes the arrival of ground forces at the pickup zone in the proper order for movement. It prescribes what actions the ground force must complete to prepare to load the aircraft. All vehicles and equipment to be lifted should be properly configured, inspected, and ready to load before the aircraft arrive at the pickup zone. Typically, ground forces arrive at the pickup zone and posture in proper chalk order before their aircraft arrive. Considerations to develop a staging plan are addressed in this section.

PREPARATION FOR LOADING

13-26. Preparations for loading are conducted in a unit assembly area or other secure location that is near the pickup zone. Before reporting to the pickup zone, units complete all preparations to successfully load the aircraft, to include—

- Completing the air-loading table or manifest. The chalk leaders verify the air-loading table to ensure it is properly completed, making changes to the manifest before arriving to the pickup zone.
- Preparing and inspecting all equipment for loading. The chalk leaders prerig all equipment to be sling loaded and ensure vehicles have the proper equipment to rig and fly. The chalk leaders inspect their loads and complete all necessary inspection records, to include DA Form 7382, *Sling Load Inspection Record*, according to TM 4-48.09.
- Conducting rehearsals for loading and off-loading the aircraft.

MOVEMENT TO PICKUP ZONE

13-27. Once units have completed preparations for loading, they begin movement to the pickup zone according to the air movement table so that the Soldiers to load arrive shortly prior to the helicopter to be loaded. This prevents congestion, preserves security, and reduces vulnerability to enemy actions on the pickup zone. To coordinate the movement of units to the pickup zone, the PZCO—

- Determines movement time of ground forces to the pickup zone.
- Specifies arrival time(s).
- Ensures that movement of units remains on schedule.

CHALK CHECK-IN AND INSPECTION

13-28. Upon arriving to the pickup zone area, the unit first checks in with the pickup zone control party at chalk check-in. The PZCO should plan adequate time for check-in based on mission variables. As a rule, the greater the number of serials in a lift, the longer it takes check-in and inspection for loading. Serials with large numbers of vehicles and equipment to be sling loaded require more time to check in.

CHALK CHECK-IN

13-29. As the unit arrives at the check-in point, loads are identified by lift-serial-chalk. Chalk leaders are briefed, and their air loading tables or manifests are inspected. The chalk leader provides one copy of the manifest to the pickup zone control party.

LOAD WEIGH-IN

13-30. The loads then are weighed with all personnel and equipment to ensure they meet the ACLs as briefed in the air mission brief. Overweight loads are sent to a designated frustrated cargo area to download equipment before being reweighed.

Chapter 13

LOAD INSPECTION

13-31. All items to be loaded are inspected according to TM 4-48.09. For emergency purposes only, the pickup zone control party may maintain a parts box for on-the-spot corrections. Units are responsible for the serviceability and corrective maintenance of their own equipment.

13-32. Loads with deficiencies are sent to a designated frustrated cargo area. Loads must remain in the frustrated area until deficiencies are corrected and the loads are inspected again. No load is allowed to leave the frustrated area without permission from the PZCO.

LOAD STAGING

13-33. Once a serial is complete, a chalk guide from the pickup zone control leads it into position on the pickup zone. Loads are staged in reverse chalk order by serial according to the pickup zone diagram.

13-34. Once the chalk is staged and in pickup zone posture, the chalk leader should brief his chalk on—
- Seating arrangement.
- Loading procedures.
- Use of safety belts.
- In-flight procedures.
- Off-loading procedures.

SLING LOAD OPERATIONS

13-35. The three phases of a sling load operation are—
- Preparation and rigging. Loads are prepared and rigged according to TM 4-48.09 or unit standard operating procedures.
- Inspection. A Pathfinder School graduate, Sling Load Inspector Certification Course graduate, or an Air Assault School graduate in the rank of specialist and above is qualified to inspect and certify each load. The individual who rigged the load cannot inspect the same load. The contents of the load are recorded on a DA Form 7382.
- Sling load operation. Trained ground crews hook up loads.

SLING LOAD UNITS

13-36. The three different elements involved in a sling load operation are the supported unit, the aviation unit, and the receiving unit. In an air assault, the supported unit and the receiving unit are the same. The responsibilities of each element are as described below.
- Support unit is responsible for—
  - Selecting, preparing, and controlling the pickup zone.
  - Requisitioning all the equipment needed for sling load operations.
  - Inspecting and maintaining all sling load equipment.
  - Providing trained ground crews for rigging and inspecting, filing inspection forms, controlling aircraft, aircraft guides, hooking up loads, and clearing the aircraft for departure.
  - Providing load dispositions and instructions to the aviation unit for the sling load equipment.
  - Verifying the load weight (to include rigging equipment).
- Aviation unit is responsible for—
  - Establishing coordination with the supported unit.
  - Advising the supported unit on load limitations.
  - Advising the supported units on the suitability of selected landing zones and pickup zones.
  - Providing assistance in the recovery and return of sling load equipment.
  - Establishing safety procedures and understanding of duties and responsibility between the flight crew and ground crew.
- Receiving unit is responsible for—

- Selecting, preparing, and controlling the landing zone.
- Providing trained ground crews to guide the aircraft and derig the loads.
- Coordinating for the control and return of the sling load equipment.
- Inspecting the rigging of back loads (sling load equipment returning to pickup zone).

SLING LOAD TEAMS

13-37. Three personnel are used for the ground crew in external load operations on the pickup zone or landing zone. They are—
- Signal person.
- Static probe person.
- Hook-up person.

13-38. The static probe person carries an electricity probe an insulated contact rod joined by a length of metallic tape or electrical wire to a ground rod. All ground crew personnel wear the following protective equipment:
- Advanced combat helmet.
- Goggles.
- Earplugs.
- Gloves.
- Sleeves rolled down and buttoned.
- Identification card and tags.

HOOK-UP SITE

13-39. The aircraft approaches the hook-up site, and the signal person guides it into position over the load. The static probe person drives the ground rod into the ground and discharges the static electricity from the aircraft by holding the contact rod, which is connected to the ground rod, to the cargo hook of the aircraft. The hook-up person then attaches the apex fitting to the aircraft cargo hook.

Note. When using a cargo hook pendant the use of a static discharge wand is not required.

RELEASE SITE

13-40. The aircraft approaches the release site, and the signal person guides it into position. The hook-up release team stands by but is not actively employed unless the slings cannot be released from the aircraft. The ground crew at the landing zone comprises one signal person and two release personnel.

GROUND CREW EMERGENCY

13-41. In an emergency, the ground crew moves to a predesignated rendezvous point identified during prior coordination with the aviation unit. Thorough preparation and rehearsal enable ground crews to react to changes to the plan and unexpected events.

This page intentionally left blank.

Glossary

The glossary lists acronyms and terms with Army or joint definitions. Where Army and joint definitions differ, (Army) precedes the definition. Terms for which FM 3-99 is the proponent are marked with an asterisk (*). The proponent manual for other terms is listed in parentheses after the definition.

SECTION I – ACRONYMS AND ABBREVIATIONS

Acronym	Definition
	A
AACG	arrival airfield control group
AADC	area air defense commander
AAGS	Army air-ground system
AAMDC	Army air missile defense command
AATF	air assault task force
AATFC	air assault task force commander
ABCT	Armored brigade combat team
ABN	airborne
ABNAFC	airborne assault force commander
ABNAF	airborne assault force
ABN IBCT	airborne Infantry brigade combat team
ABNTF	airborne task force
ABNTFC	airborne task force commander
ACL	allowable cargo load
A/DACG	arrival/departure airfield control group
ADAM	air defense airspace management
ADP	Army doctrine publication
ADRP	Army doctrine reference publication
AGL	above ground level
ALCC	airlift control center
ALCE	airlift control element
AMCM	air mission coordination meeting
AMD	air missile defense
AMPS	Aviation Mission Planning System
ANGLICO	Air-naval gunfire liaison company
APOD	aerial port of debarkation
ATP	Army techniques publication
ATTP	Army tactics, techniques, and procedures
AWACS	Airborne Warning and Control System

Glossary

Acronym	Definition

B

BAE	brigade aviation element
BAO	brigade aviation officer
BCT	brigade combat team
BN	battalion

C

CAN	combat aviation network
CAOC	combat air operations center
CARP	computed air release point
CATF	commander, amphibious task force
CBRN	chemical, biological, radiological, and nuclear
CCIR	commander's critical information requirement
CDRJSOTF	commander, joint special operation task force
CH	cargo helicopter
CLF	commander, landing force
CONOPS	concept of operations

D

DACG	departure airfield control group
DACO	departure airfield control officer
DA	Department of the Army
DD	Department of Defense form
DZ	drop zone

E

EDRE	emergency deployment readiness exercise
EPLRS	Enhanced Position Location Reporting System

F

FAC (A)	forward air controller (airborne)
FARP	forward arming and refueling point
FASCAM	field artillery scatterable mines
FBCB2	Force XXI Battle Command-Brigade and Below
FM	field manual
FRAGORD	fragmentary order
FSCM	fire support coordination measures
FSC	forward support company
FRIES	Fast-Insertion/Extraction System

Acronym	Definition
	G
GMRS	Ground Marked Relief System
	H
HAHO	high-altitude high-opening parachute
HALO	high-altitude low-opening parachute
HEPI	heavy equipment point of impact
HF	high frequency
HIDACZ	high-density airspace control zone
HMMWV	high-mobility multipurpose wheeled vehicle
	I
IBCT	Infantry brigade combat team
ICODES	Integrated Computerized Deployment System
ISB	intermediate staging base
	J
J-2	intelligence directorate of a joint staff
J-3	operations directorate of a joint staff
JACC/CP	joint airborne communication center/command post
JFACC	joint force air component commander
JFLCC	joint force land component commander
JFMCC	joint force maritime component commander
JIOC	joint intelligence operations center
JIPOE	joint intelligence preparation of the operational environment
JOPP	joint operation planning process
JP	joint publication
JPADS	Joint Precision Airdrop System
JSOA	joint special operations area
JSTARS	Joint Surveillance Target Attack Radar System
JTAC	joint terminal attack controller
	L
LACC	loading area control center
LZ	landing zone
LRSC	long-range surveillance company
	M
MDMP	military decisionmaking process
METT-TC	mission, enemy, terrain and weather, troops and

Acronym	Definition
	support available-time available, and civil considerations
MILDEC	military deception
MMEE	minimum mission essential equipment

N

NCO	noncommissioned officer
NVG	night vision goggles

O

OAKOC	observation and fields of fire, avenues of approach, key terrain, obstacles, cover and concealment
OPORD	operation order
OPSEC	operations security

P

POL	petroleum, oils, and lubricants
PZ	pickup zone
PZCO	pickup zone control officer
PZNCOIC	pickup zone noncommissioned officer-in-charge

R

RADC	regional air defense commander
RATELO	radiotelephone operator
RDSP	rapid decisionmaking and synchronization process
RSOI	reception, staging, onward movement, integration

S

S-1	personnel staff officer
S-2	intelligence staff officer
S-3	operations staff officer
S-4	logistics staff officer
S-6	signal staff officer
S-9	civil affairs operations staff officer
SATCOM	satellite communication
SBCT	Stryker brigade combat team
SDAC	sector air defense commander
SEAD	suppression of enemy air defenses
SIPRNET	Secret Internet Protocol Router Network
SLOC	sea lines of communication
SOP	standard operating procedure

Glossary

Acronym	Definition
SPIES	Special Patrol Insertion/Extraction System
	T
TACP	tactical air control party
TACSAT	tactical satellite
TAIS	Tactical Airspace Integration System
TM	technical manual
	U
UH	utility helicopter
UHF	ultra-high frequency
USAF	United States Air Force
USMC	United States Marine Corps
USN	United States Navy
U.S.	United States
	V
VHF	very high frequency
	W
WARNORD	warning order

SECTION II – TERMS

air assault

   The movement of friendly assault forces by rotary-wing aircraft to engage and destroy enemy forces or to seize and hold key terrain. (JP 3-18)

air assault force

   A force composed primarily of ground and rotary-wing air units organized, equipped, and trained for air assault operations. (JP 3-18)

air assault operation

   An operation in which assault forces, using the mobility of rotary-wing assets and the total integration of available firepower, maneuver under the control of a ground or air maneuver commander to engage enemy forces or to seize and hold key terrain. (JP 3-18)

air movement

   Air transport of units, personnel, supplies, and equipment including airdrops and air landings. (JP 3-17)

airborne assault

   The use of airborne forces to parachute into an objective area to attack and eliminate armed resistance and secure designated objectives. (JP 3-18)

airborne operation

   An operation involving the air movement into an objective area of combat forces and their logistic support for execution of a tactical, operational, or strategic mission. (JP 3-18)

Glossary

airfield

An area prepared for the accommodation (including any buildings, installations, and equipment), landing, takeoff of aircraft. (JP 3-17)

airhead

A designated area in a hostile or potentially hostile operational area that, when seized and held, ensures the continuous air landing of troops and materiel and provides the maneuver space necessary for projected operations. (JP 3-18)

airhead line

A line denoting the limits of the objective area for an airborne assault. (JP 3-18)

airspace coordinating measures

Measures employed to facilitate the efficient use of airspace to accomplish missions and simultaneously provide safeguards for friendly forces. (JP 3-52)

airspace coordination area

A three-dimensional block of airspace in a target area, established by the appropriate ground commander, in which friendly aircraft are reasonably safe from friendly surface fires. The airspace coordination area may be formal or informal. (JP 3-09.3)

*assault echelon

(Army) The element of a force that is secheduled for initial assault on the objective area.

boundary

A line that delineates surface areas for the purpose of facilitating coordination and deconfliction of operations between adjacent units, formations, or areas. (JP 3-0)

civil considerations

The influence of manmade infrastructure, civilian institutions, and activities of the civilian leaders, populations, and organizations within an area of operations on the conduct of military operations. (ADRP 5-0)

close air support

Air action by fixed- and rotary-wing aircraft against hostile targets that are in close proximity to friendly forces and that require detailed integration of each air mission with the fire and movement of those forces. (JP 3-0)

close combat attack

A coordinated attack by Army attack reconnaissance aircraft (manned and unmanned) against enemy forces that are in close proximity to friendly forces. The close combat attack is not synonymous with close air support flown by joint aircraft. Terminal control from ground units or controllers is not due to the capabilities of the aircraft and the enhanced situational understanding of the aircrew. (FM 3-04.126)

combat identification

The process of attaining an accurate characterization of detected objects in the operational environment to support an engagement decision. (JP 3-09)

command group

The commander and selected staff members who assist the commander in controlling operations away from a command post. (FM 6-0)

commander's intent

A clear and concise expression of the purpose of the operation and the desired military end state that supports mission command, provides focus to the staff, and helps subordinate and supporting commanders act to achieve the commander's desired results without further orders, even when the operation does not unfold as planned. (JP 3-0)

concept of operations

A statement that directs the manner in which subordinate units cooperate to accomplish the mission and establishes the sequence of actions the force will use to achieve the end state. (ADRP 5-0)

concealment

The protection from observation or surveillance. (ADRP 1-02)

control measure

A means of regulating forces or warfighting functions. (ADRP 6-0)

cover

Protection from the effects of fires. (ADRP 1-02)

D-day

The unnamed day on which a particular operation commences or is to commence. (JP 3-02)

decisive operation

The operation that directly accomplishes the mission. (ADRP 3-0)

electromagnetic operational environment

The background electromagnetic environment and the friendly, neutral, and adversarial electromagnetic order of battle within the electromagnetic area of influence associated with a given operational area. (JP 6-01)

electromagnetic spectrum management

The planning, coordinating, and managing use of the electromagnetic spectrum through operational, engineering, and administrative procedures. (JP 6-01)

fire support coordination measure

A measure employed by commanders to facilitate the rapid engagement of targets and simultaneously provide safeguards for friendly forces. (JP 3-0)

*follow-on echelon

(Army) Those additional forces moved into the objective area after the assault echelon.

forcible entry

The seizing and holding of a military lodgment in the face of armed opposition. (JP 3-18)

H-hour

The specific hour on D-day at which a particular operation commences. (JP 3-02)

information environment

The aggregate of individuals, organizations, and systems that collect, process, disseminate, or act on information. (JP 3-13)

information operations

The integrated employment, during military operations, of information-related capabilities in concert with other lines of operation to influence, disrupt, corrupt, or usurp the decisionmaking of adversaries and potential adversaries while protecting our own. (JP 3-13)

information superiority

The operational advantage derived from the ability to collect, process, and disseminate an uninterrupted flow of information while exploiting or denying an adversary's ability to do the same. (JP 3-13)

information system

Equipment that collect, process, store, display, and disseminate information. This includes computers—hardware and software—and communications, as well as policies and procedures for their use. (ADP 6-0)

intermediate staging base

A tailorable, temporary location used for staging forces, sustainment and/or extraction into and out of an operational area. (JP 3-35)

lodgment
> A designated area in a hostile or potentially hostile operational area that, when seized and held, makes the continuous landing of troops and materiel possible and provides maneuver space for subsequent operations. (JP 3-18)

main effort
> A designated subordinate unit whose mission at a given point in time is most critical to overall mission success. (ADRP 3-0)

marshalling
> The process by which units participating in an amphibious or airborne operation group together or assemble when feasible or move to temporary camps in the vicinity of embarkation points, complete preparations for combat, or prepare for loading. (JP 3-17)

mission command
> The exercise of authority and direction by the commander using mission orders to enable disciplined initiative within the commander's intent to empower agile and adaptive leaders in the conduct of unified land operations. (ADP 6-0)

*N-hour
> The time a unit is notified to assemble its personnel and begin the deployment sequence.

*N-hour sequence
> Starts the reverse planning necessary after notification to have the first assault aircraft en route to the objective area for commencement of the parachute assault according to the order for execution.

obstacles
> Any natural or man-made obstruction designed or employed to disrupt, fix, turn, or block the movement of an opposing force, and to impose additional losses in personnel, time, and equipment on the opposing force. (JP 3-15)

P-hour
> The specific hour on D-day at which a parachute assault commences with the exit of the first Soldier from an aircraft over a designated drop zone. P-hour may or may not coincide with H-hour. (FM 6-0)

phase
> A planning and execution tool used to divide an operation in duration or activity. (ADRP 3-0)

planned target
> A target that is known to exist in the operational environment, upon which actions are planned using deliberate targeting, creating effects which support the commander's objectives. (JP 3-60)

*rear echelon
> The echelon containing those elements of the force that are not required in the objective area.

reconnaissance
> A mission undertaken to obtain, by visual observation or other detection methods, information about the activities and resources of an enemy or adversary, or to secure data concerning the meteorological, hydrographic, or geographic characteristics of a particular area. (JP 2-0)

reorganization
> All measures taken by the commander to maintain unit combat effectiveness or return it to a specified level of combat capability. (FM 3-90-1)

restricted operations area
> Airspace of defined dimensions, designated by the airspace control authority, in response to specific operational situations/requirements within which the operation of one or more airspace users is restricted. (JP 3-52)

Glossary

security operations
> Those operations undertaken by a commander to provide early and accurate warning of enemy operations, to provide the force being protected with time and maneuver space within which to react to the enemy, and to develop the situation to allow the commander to effectively use the protected force. (ADRP 3-90)

shaping operation
> An operation that establishes conditions for the decisive operation through effects on the enemy, other actors, and the terrain. (ADRP 3-0)

supporting effort
> A designated subordinate unit with a mission that supports the success of the main effort. (ADRP 3-0)

surveillance
> The systematic observation of aerospace, surface, or subsurface areas, places, persons, or things, by visual, aural, electronic, photographic, or other means. (JP 3-0)

sustaining opeation
> An operation at any echelon that enables the decisive operation or shaping operation by generating and maintaining combat power. (ADRP 3-0)

task organization
> A temporary grouping of forces designed to accomplish a particular mission. (ADRP 5-0)

vertical envelopment
> A tactical maneuver in which troops that are air-dropped, air-landed, or inserted via air assault, attack the rear and flanks of a force, in effect cutting off or encircling the force. (JP 3-18)

*X-hour
> The unspecified time that commences unit notification for planning and deployment preparation in support of potential contingency operations that do not involve rapid, short notice deployment.

*X-hour sequence
> An extended sequence of events initiated by X-hour that allow a unit to focus on planning for a potential contingency operation, to include preparation for deployment.

This page intentionally left blank.

References

REQUIRED PUBLICATIONS
These documents must be available to the intended user of this publication.
ADRP 1-02. *Terms and Military Symbols.* 24 September 2013.
JP 1-02. *Department of Defense Dictionary of Military and Associated Terms.* 8 November 2010.

JOINT PUBLICATIONS
Most joint publications are available online:
<http://www.dtic.mil/doctrine/new_pubs/jointpub.htm.>
JP 2-0. *Joint Intelligence.* 22 October 2013.
JP 2-01. *Joint and National Intelligence Support to Military Operations.* 5 January 2012.
JP 2-01.3. *Joint Intelligence Preparation of the Operational Environment.* 21 May 2014.
JP 3-0. *Joint Operations.* 11 August 2011.
JP 3-01. *Countering Air and Missile Threats.* 23 March 2012.
JP 3-02. *Amphibious Operations.* 18 July 2014.
JP 3-03. *Joint Interdiction.* 14 October 2011.
JP 3-05. *Special Operations.* 16 July 2014.
JP 3-09. *Joint Fire Support.* 12 December 2014.
JP 3-09.3. *Close Air Support.* 25 November 2014.
JP 3-11. *Operations in Chemical, Biological, Radiological, and Nuclear Environments.* 4 October 2013.
JP 3-13. *Information Operations.* 27 November 2012.
JP 3-13.3. *Operations Security.* 4 January 2012.
JP 3-13.4. *Military Deception.* 26 January 2012.
JP 3-14. *Space Operations.* 29 May 2013.
JP 3-15. *Barriers, Obstacles, and Mine Warfare for Joint Operations.* 17 June 2011.
JP 3-17. *Air Mobility Operations.* 30 September 2013.
JP 3-18. *Joint Forcible Entry Operations.* 27 November 2012.
JP 3-30. *Command and Control of Joint Air Operations.* 10 February 2014.
JP 3-32. *Command and Control for Joint Maritime Operations.* 07 August 2013.
JP 3-35. *Deployment and Redeployment Operations.* 31 January 2013.
JP 3-40. *Countering Weapons of Mass Destruction.* 31 October 2014.
JP 3-52. *Joint Airspace Control.* 13 November 2014.
JP 3-59. *Meteorological and Oceanographic Operations.* 7 December 2012.
JP 3-60. *Joint Targeting.* 31 January 2013.
JP 5-0. *Joint Operation Planning.* 11 August 2011.
JP 6-0. *Joint Communications System.* 10 June 2010.
JP 6-01. *Joint Electromagnetic Spectrum Management Operations.* 20 March 2012.

ARMY PUBLICATIONS
Most army doctrinal publications are available online:
https://armypubs.us.army.mil/doctrine/Active_FM.html.
ADP 3-0. *Unified Land Operations.* 10 October 2011.

References

ADP 5-0. *The Operations Process.* 17 May 2012

ADP 6-0. *Mission Command.* 17 May 2012.

ADRP 2-0. *Intelligence.* 31 August 2012.

ADRP 3-0. *Unified Land Operations.* 16 May 2012.

ADRP 3-05. *Special Operations.* 31 August 2012.

ADRP 3-37. *Protection.* 31 August 2012.

ADRP 3-90. *Offense and Defense.* 31 August 2012.

ADRP 5-0. *The Operations Process.* 17 May 2012.

ADRP 6-0. *Mission Command.* 17 May 2012.

ATP 1-02.1. Multi-Service Tactics, Techniques, and Procedures for Multi-Service Brevity Codes. 23 October 2014.

ATP 2-01. *Plan Requirements and Assess Collection.* 19 August 2014.

ATP 2-01.3. *Intelligence Preparation of the Battlefield/Battlespace.* 10 November 2014.

ATP 3-01.4. *Multiservice Tactics, Techniques, and Procedures for Joint Suppression of Enemy Air Defense (J-SEAD).* 19 July 2013.

ATP 3-09.32. *JFIRE Multiservice Tactics, Techniques, and Procedures for the Joint Application of Firepower.* 30 November 2012.

ATP 3-36. *Electronic Warfare Techniques.* 16 December 2014.

ATP 3-60.1. *Multi-Service Tactics, Techniques, and Procedures for Dynamic Targeting.* 7 May 2012.

ATTP 3-18.04. *Special Forces Special Reconnaissance Tactics, Techniques, and Procedures.* 5 January 2011.

ATP 4-02.2. *Medical Evacuation.* 12 August 2014.

ATP 5-19. *Risk Management.* 14 April 2014.

FM 2-0. *Intelligence Operations.* 15 April 2014.

FM 2-22.3. *Human Intelligence Collector Operations.* 6 September 2006.

FM 3-01. *United States Army Air and Missile Defense Operations.* 15 April 2014.

FM 3-04.111. *Aviation Brigades.* 7 December 2007.

FM 3-04.113. *Utility and Cargo Helicopter Operations.* 7 December 2007.

FM 3-04.126. *Attack Reconnaissance Helicopter Operations.* 16 February 2007.

FM 3-04.155. *Army Unmanned Aircraft System.* 29 July 2009.

FM 3-05. *Army Special Operations.* 9 January 2014.

FM 3-05.210. *Special Forces Air Operations.* 27 February 2009.

FM 3-09.*Field Artillery Operations and Fire Support.* 4 April 2014.

FM 3-14. *Army Space Operations.* 19 August 2014.

FM 3-16. *The Army in Multinational Operations.* 8 April 2014.

FM 3-17.2. *Multi-Service Tactics, Techniques and Procedures for Airfield Opening.* 15 May 2007.

FM 3-21.8. *The Infantry Rifle Platoon and Squad.* 28 March 2007.

FM 3-21.10. *The Infantry Rifle Company.* 27 July 2006.

FM 3-21.20. *The Infantry Battalion.* 13 December 2006.

FM 3-21.38. *Pathfinder Operations.* 25 April 2006.

FM 3-35. *Army Deployment and Redeployment.* 21 April 2010.

FM 3-38. *Cyber Electromagnetic Activities.* 12 February 2014.

FM 3-52. *Airspace Control.* 8 February 2013.

FM 3-55. *Information Collection.* 3 May 2013.

FM 3-55.93. *Long-Range Surveillance Unit Operations.* 23 June 2009.

FM 3-60. *The Targeting Process.* 26 November 2010.
FM 3-90-1. *Offense and Defense Volume 1.* 22 March 2013.
FM 3-90-2. *Reconnaissance, Security, and Tactical Enabling Tasks Volume 2.* 22 March 2013.
FM 3-90.6. *Brigade Combat Team.* 14 September 2010.
FM 3-94. *Theater Army, Corps, and Division Operations.* 21 April 2014.
FM 6-0. *Commander and Staff Organization and Operations.* 5 May 2014.
FM 6-02. *Signal Support to Operations.* 22 January 2014.
FM 6-02.53. *Tactical Radio Operations.* 5 August 2009.
FM 6-05. *CF-SOF Multi-Service Tactics, Techniques, and Procedures for Conventional Forces and Special Operations Forces Integration, Interoperability, and Interdependence.* 13 March 2014.
FM 27-10. *The Law of Land Warfare.* 18 July 1956.
TC 3-21.220. *Static Line Parachuting Techniques and Training.* 28 April 2014.
TM 4-48.09. *Multiservice Helicopter Sling Load Basic Operations and Equipment.* 23 July 2012.

WEBSITES

Army Knowledge Online, https://armypubs.us.army.mil/doctrine/index.html.

Army Publishing Directorate, http://www.apd.army.mil/.

Central Army Registry (CAR) on the Army Training Network (ATN), https://atiam.train.army.mil. CAC or AKO login required.

PRESCRIBED FORMS

There are no prescribed forms for this publication.

REFERENCED FORMS

Forms are available on the APD Web site (www.apd.army.mil)

DA Form 2028. *Recommended Changes to Publications and Blank Forms.*

DA Form 7382. *Sling Load Inspection Record.*

DD Form 1387-2. *Special Handling Data/Certification*

DD Form 2131. *Passenger Manifest.*

This page intentionally left blank.

Index

AACG, 7-9
AADC, 1-11, 1-12, 1-13, 3-10
AATF, 1-10, 1-26, 8-1, 8-2, 8-4, 8-5, 8-6, 8-7, 8-11, 8-13, 8-14, 8-15, 8-16, 8-17, 9-1, 9-2, 9-3, 9-4, 9-5, 9-6, 9-7, 9-8, 9-9, 9-10, 9-11, 9-13, 9-14, 10-1, 10-2, 10-3, 10-5, 10-7, 10-8, 10-9, 10-11, 10-12, 11-3, 12-1, 12-4, 12-5, 12-6, 12-7, 13-1, 13-3
AATFC, 1-10, 1-26, 8-1, 8-11, 8-12, 8-13, 8-14, 8-15, 8-16, 8-17, 9-1, 9-2, 9-3, 9-5, 9-6, 9-7, 9-8, 9-10, 9-11, 9-13, 9-14, 10-2, 10-3, 10-4, 10-7, 10-8, 10-12, 11-1, 11-2, 11-3, 11-4, 11-5, 12-1, 12-6, 13-1, 13-2, 13-3
    air assault task force commander, 1-25, 9-8, 9-10, 10-4, 10-8, 10-11, 11-2, 11-3, 13-2, 13-6
ABNAF, 1-22, 1-23, 2-1, 2-2, 2-3, 2-4, 2-5, 2-6, 3-1, 3-2, 3-3, 3-6, 3-7, 3-8, 3-9, 3-10, 3-12, 4-1, 4-2, 4-3, 4-13, 4-15, 4-16, 4-20, 5-1, 5-10, 6-1, 6-9, 7-1, 7-4
ABNAFC, 1-9, 1-23, 2-2, 3-1, 3-2, 3-3, 3-8, 3-12, 4-1, 4-2, 4-3, 4-8, 4-13, 4-15, 4-16, 5-1, 5-2, 5-3, 5-10, 6-9, 13-6
ABNTFC, 1-9, 1-23, 1-24, 2-2, 3-2, 3-3, 3-8, 4-16
ACLs, 6-2, 7-9, 9-6, 13-8
ACMs, 1-12, 3-9, 9-12
ADAM/BAE, 1-11, 4-13, 8-8, 8-14, 8-15, 9-2, 9-5, 10-8
aerial casualty evacuation, 9-16
Air assault task force commander, 1-10
air defense airspace management/brigade aviation element, 4-13, 8-14, 9-2
air liaison officer, 1-11, 4-15, 4-16, 5-4, 8-13, 8-15, 10-11
air mission coordination meeting, 9-5, 9-7, 9-8, 13-1

airborne assault force, 1-9, 1-22, 2-1, 2-2, 2-7, 7-1
airborne assault force commander, 1-23
airborne task force commander, 1-9, 1-23
Airborne Warning and Control System, 1-13, 2-6
airspace coordinating measures, 1-9, 1-12, 3-9, 3-11, 8-7, 8-14, 9-12
ALCE, 6-6, 6-9, 7-9
ALO, 8-15
AMCM, 9-5, 9-6, 9-7, 13-1
assembly aids, 5-18, 5-19, 5-21, 5-22, 5-24, 5-25, 5-26
aviation liaison officer, 4-13, 8-8, 8-15, 9-2, 10-8, 11-1, 12-1, 12-7, 13-3
AWACS, 1-13, 2-6, 2-7
casualty backhaul, 9-15
casualty evacuation, 9-14
CATF
    commander, amphibious task force, 1-10
CCA 5-Line attack brief, 4-13, 4-14, 10-9
commander of the air assault task force, 1-10
commander, amphibious task force, 1-8
commander, joint special operations task force, 1-8
concept of operations, 1-5, 1-12, 3-3, 8-11, 9-2
CONOPS
    concept of operations, 1-5, 1-9, 1-10, 1-18, 1-21, 3-3, 8-11, 9-2, 9-5, 10-2
counterair, 1-2, 1-12, 1-13, 5-11, 5-14, 5-15, 6-2
Counterair, 5-11
DACG, 6-6, 6-9, 7-9
DACO, 7-8
drop zone, 1-12, 1-24, 2-2, 2-4, 2-6, 3-8, 4-9, 4-11, 4-12, 4-13, 4-15, 5-1, 5-2, 5-3, 5-4, 5-5, 5-6, 5-7, 5-8, 5-9, 5-10, 5-11, 5-17, 5-18, 5-19, 5-20, 5-21, 5-23, 5-24, 5-25, 5-26, 6-1, 6-3, 6-4, 6-5, 6-6, 6-8, 6-9

drop zones, 2-2, 2-4, 3-2, 3-9, 4-3, 4-4, 4-7, 4-11, 4-12, 4-16, 5-1, 5-2, 5-5, 5-6, 5-7, 5-8, 5-9, 5-10, 5-16, 5-25, 6-1, 6-2, 7-9
elements of air movement plan, 6-1
fire support, 3-3, 3-9, 3-10, 5-11, 5-12, 5-13, 5-14, 6-3, 8-8, 8-17, 10-8, 12-4
Fire support, 5-10
fire support coordination measures, 1-9, 3-10, 8-7, 9-11
FSCMs, 1-9, 3-9, 3-10, 9-12
    fire support coordination measures, 3-9
high-density air control zone, 1-12
intelligence preparation of the battlefield, 1-15, 1-16, 3-7
intermediate staging base, 5-12
IPB, 1-15, 3-7
ISB, 1-14, 4-20
    intermediate staging base, 1-7, 2-7, 4-20, 4-22, 5-17, 6-2, 7-3, 7-8
JACC/CP, 2-6, 2-7, 2-8, 5-12
JACCEs, 1-11
JFACC, 1-9, 1-10, 1-11, 1-12, 1-13
JFC
    joint force commander, 1-4, 1-5
JFLCC, 1-9
    joint force land commander, 1-8, 1-13
JFMCC, 1-10
JIOC, 1-15
joint air component coordination elements, 1-11
Joint Airborne Communications Center/Command Post, 2-6, 2-7
joint force commander, 1-2, 1-4, 1-5, 1-6, 1-8, 1-9, 1-10,

Index

1-11, 1-13, 1-19, 1-23, 3-2, 3-10, 5-16
joint force land component commander, 1-8
Joint force maritime component commander, 1-10
joint operation planning process, 1-14
Joint Precision Airdrop System, 5-4
joint special operations task force, 3-8
Joint Surveillance Target Attack Radar System, 2-6, 2-7
joint task force joint intelligence operations center, 1-15
joint terminal attack controllers, 4-15
JOPP, 1-14
JPADS, 5-4
JSOA
    joint special operations area, 1-8
JSOTF, 1-10, 3-8
JSTARS, 2-6, 2-7
JTACs, 4-15
LACC, 7-6

landing zone, 1-12, 1-26, 3-3, 4-9, 4-13, 5-1, 5-5, 5-19, 5-26, 6-1, 6-8, 6-9, 8-4, 8-8, 8-9, 8-10, 8-14, 9-3, 9-7, 9-8, 9-9, 9-10, 9-14, 10-1, 10-3, 10-10, 10-11, 10-13, 11-1, 11-2, 11-3, 11-4, 11-5, 11-6, 11-7, 11-8, 11-10, 11-11, 11-12, 11-13, 12-1, 12-2, 12-4, 12-6, 12-7, 12-8, 13-1, 13-5, 13-6, 13-9, 13-10
marshalling area, 2-1, 3-4, 4-1, 7-2, 7-4, 7-5, 7-6, 7-7, 7-8, 7-9
marshalling plan, 3-1, 3-3, 3-4, 3-7, 4-1, 7-1, 7-4, 7-9
MDMP, 3-1, 3-4, 3-5, 9-1, 9-3, 9-4, 9-5, 9-6, 12-5
medical evacuation, 9-14
    landing zone, 9-16
    planning, 9-14
N-hour, 3-6, 7-1
operational environment, 1-1, 1-2, 1-3, 1-4, 1-7, 1-9, 1-14, 1-15, 1-16, 8-2
PZCO, 8-15, 13-2, 13-3, 13-7, 13-8
PZNCOIC, 13-2
RDSP, 9-5
regional (or sector) air defense commander, 1-8

remote marshalling base, 2-3, 4-20, 4-21, 6-2
sea lines of communications, 1-3
SEAD, 1-25, 2-6, 3-12, 4-12, 5-15, 9-9, 12-4, 12-5, 12-6
special patrol infiltration and exfiltration system, 13-1
suppression of enemy air defenses, 2-6, 6-2, 9-13, 10-12, 12-4
TACC, 2-7
TACP, 3-9, 3-10, 4-15, 4-16, 5-12, 5-13, 8-15, 10-11
TACSAT, 1-24, 2-7, 7-12, 8-16, 8-17
tactical air control party, 1-11, 3-9, 4-15, 5-12
tactical airlift control center, 2-7
Tactical Airspace Integration System, 8-14
TAIS, 8-14
the fast-rope insertion and extraction system, 11-6
UAS, 4-16, 4-17, 8-8, 8-14, 9-13, 9-14, 10-3, 10-9, 10-12, 11-3, 12-6
X-hour, 3-6

FM 3-99
6 March 2015

By Order of the Secretary of the Army:

RAYMOND T. ODIERNO
General, United States Army
Chief of Staff

Official:

GERALD B. O'KEEFE
Administrative Assistant to the
Secretary of the Army
1504001

DISTRIBUTION:
Active Army, Army National Guard, and U.S. Army Reserve: Distributed in electronic media only (EMO).

Made in United States
Orlando, FL
29 May 2025